ISBN 0-8373-6601-1

DANTES– 1

Rudman's Question and Answers on the

DANTES SUBJECT STANDARDIZED TESTS

Subject Examination In . . .

ASTRONOMY

Questions and Answers

NLC
NATIONAL LEARNING CORPORATION
212 MICHAEL DRIVE, SYOSSET, NEW YORK 11791 (516) 921-8888

National Learning Corporation

212 Michael Drive, Syosset, New York 11791
(516) 921-8888

PRINTED IN THE UNITED STATES OF AMERICA

PASSBOOK®

NOTICE

This book is *SOLELY* intended for, is sold *ONLY* to, and its use is *RESTRICTED* to *individual*, bona fide applicants or candidates who qualify by virtue of having seriously filed applications for appropriate license, certificate, professional and/or promotional advancement, higher school matriculation, scholarship, or other legitimate requirements of educational and/or governmental authorities.

This book is *NOT* intended for use, class instruction, tutoring, training, duplication, copying, reprinting, excerption, or adaptation, etc., by:

(1) Other Publishers

(2) Proprietors and/or Instructors of "Coaching" and/or Preparatory Courses

(3) Personnel and/or Training Divisions of commercial, industrial, and governmental organizations

(4) Schools, colleges, or universities and/or their departments and staffs, including teachers and other personnel

(5) Testing Agencies or Bureaus

(6) Study groups which seek by the purchase of a single volume to copy and/or duplicate and/or adapt this material for use by the group as a whole without having purchased individual volumes for each of the members of the group

(7) Et al.

Such persons would be in violation of appropriate Federal and State statutes.

PROVISION OF LICENSING AGREEMENTS. — Recognized educational commercial, industrial, and governmental institutions and organizations, and others legitimately engaged in educational pursuits, including training, testing, and measurement activities, may address a request for a licensing agreement to the copyright owners, who will determine whether, and under what conditions, including fees and charges, the materials in this book may be used by them. In other words, a licensing facility *exists* for the legitimate use of the material in this book on other than an individual basis. However, it is asseverated and affirmed here that the materials in this book *CANNOT* be used without the receipt of the express permission of such a licensing agreement from the Publishers.

NATIONAL LEARNING CORPORATION
212 Michael Drive
Syosset, New York 11791

Inquiries re licensing agreements should be addressed to:
The President
National Learning Corporation
212 Michael Drive
Syosset, New York 11791

PASSBOOK SERIES®

THE *PASSBOOK SERIES®* has been created to prepare applicants and candidates for the ultimate academic battlefield—the examination room.

At some time in our lives, each and every one of us may be required to take an examination—for validation, matriculation, admission, qualification, registration, certification, or licensure.

Based on the assumption that every applicant or candidate has met the basic formal educational standards, has taken the required number of courses, and read the necessary texts, the *PASSBOOK SERIES®* furnishes the one special preparation which may assure passing with confidence, instead of failing with insecurity. Examination questions—together with answers—are furnished as the basic vehicle for study so that the mysteries of the examination and its compounding difficulties may be eliminated or diminished by a sure method.

This book is meant to help you pass your examination provided that you qualify and are serious in your objective.

The entire field is reviewed through the huge store of content information which is succinctly presented through a provocative and challenging approach—the question-and-answer method.

A climate of success is established by furnishing the correct answers at the end of each test.

You soon learn to recognize types of questions, forms of questions, and patterns of questioning. You may even begin to anticipate expected outcomes.

You perceive that many questions are repeated or adapted so that you gain acute insights, which may enable you to score many sure points.

You learn how to confront new questions, or types of questions, and to attack them confidently and work out the correct answers.

You note objectives and emphases, and recognize pitfalls and dangers, so that you may make positive educational adjustments.

Moreover, you are kept fully informed in relation to new concepts, methods, practices, and directions in the field.

You discover that you are actually taking the examination all the time: you are preparing for the examination by "taking" an examination, not by reading extraneous and/or supererogatory textbooks.

In short, this PASSBOOK®, used directedly, should be an important factor in helping you to pass your test.

ASTRONOMY
EXAMINATION SECTION

CONTENTS

ASTRONOMY
EXAMINATION SECTION

CONTENTS

ASTRONOMY
EXAMINATION SECTION

CONTENTS

ASTRONOMY
EXAMINATION SECTION

CONTENTS

BASIC FUNDAMENTALS OF ASTRONOMY

CONTENTS

LUNAR SCIENCE GLOSSARY

DANTES Subject Standardized Tests

INTRODUCTION

The DANTES (Defense Activity for Non-Traditional Education Support) subject standardized tests are comprehensive college and graduate level examinations given by the Armed Forces, colleges and graduate schools as end-of-subject course evaluation final examinations or to obtain college equivalency credits in the various subject areas tested.

Active military personnel may take CLEP (College-Level Examination Program) examinations without charge through the DANTES program. Information about DANTES and CLEP may be obtained from the Test Control Officers at military installations, and from Educational Testing Service in Princeton, New Jersey.

The DANTES Examination Program enables students to obtain college credit for what they have learned on the job, through self-study, personal interest, correspondence courses or by any other means. It is used by colleges and universities to award college credit to students who demonstrate that they know as much as students completing an equivalent college course. It is a cost-efficient, time-saving way for students to use their knowledge to accomplish their educational goals.

Most schools accept the American Council on Education (ACE) recommendations for the minimum score required and the amount of credit awarded, but not all schools do. Be sure to check the policy regarding the score level required for credit and the number of credits to be awarded.

Not all tests are accepted by all institutions. Even when a test is accepted by an institution, it may not be acceptable for every program at that institution. Before considering testing, ascertain the acceptability of a specific test for a particular course.

Colleges and universities that administer DANTES tests may administer them to any applicant -- or they may administer the tests only to students registered at their institution. Decisions about who will be allowed to test are made by the school. Students should contact the test center to determine current policies and schedules for DANTES testing.

Colleges and universities authorized to administer DANTES tests usually do so throughout the calendar year. Each school sets its own fee for test administration and establishes its own testing schedule. Contact the representative at the administering school directly to make arrangements for testing.

Your Experience Can Earn You Credit.

It's a fact. People learn in many ways . . . through hobbies . . . interests . . . reading . . . work. The DANTES Program is a way for you to "cash in" on what you've learned from *your* experiences.

Take advantage of DANTES tests to earn college credit for what you know. More than 30 subjects are offered. There are bound to be areas where your knowledge can pay off academically.

Here are just a few of the program's benefits for you:

- ***Time and Money Saver***
  You don't have to sit through classes and pay for courses when you already know most of what's being taught and can learn the rest yourself.
- ***Credits Widely Accepted***
  Hundreds of colleges nationwide now give credit for DANTES tests. And the list is growing!
- ***Quicker Route to the Degree***
  Your dreams are a little closer than you thought possible. With DANTES.
- ***Start Fast***
  No need to wait to take advanced courses. Passing DANTES tests gives you a real head start.

Sound Good? Here's How to Get Going!

Check with your college to see if it administers the DANTES program. If the college doesn't but would like to, ask that someone **call collect** (609-951-1756) or write to the DANTES Program Office (see address below).

If you find that your college *accepts* DANTES credits, but does not *administer* the tests, here's what you do —

- Write to the DANTES Program Office for a list of colleges that administer the test. Chances are there's one nearby.
- Call any of the ETS Regional Offices (Tucker, GA; Oakland, CA; Evanston, IL; Phoenix, AZ; and Princeton, NJ. DANTES tests are given at all of those locations.
- In the military or in the reserves? DoD employees and family members may be able to take DANTES tests at a Military Education Office or Navy Campus Office.

Check Out the DANTES Test List.

→

Match the titles with the things you have learned. Consider how much college credit you might gain by taking some DANTES tests.

Your college should have fact sheets with more information about each DANTES test. If not, please write to us directly, or e-mail us at dantes@chauncey.com. Tell us which subjects interest you.

DANTES Program Office
Educational Testing Service
Princeton, NJ 08541-0001

MATHEMATICS

Introductory College Algebra
- Functions and graphs
- Fundamental algebraic operations
- Linear equations, inequalities, word problems
- Radicals and exponents

Principles of Statistics
- Descriptive statistics
- Correlation, regression, probability
- Chance models, sampling, test of significance

SOCIAL SCIENCE

Art of the Western World
- Greece and Rome
- Romanesque and Gothic
- Renaissance
- Baroque, rococo, neoclassicism, romanticism
- Twentieth century

The Civil War and Reconstruction
- Causes of the war
- Political situations and the war between 1861 and 1865
- Reconstruction

Contemporary Western Europe
- Historical background
- Political systems
- Societal issues
- Integration

Criminal Justice
- Criminal behavior
- Police and criminal justice system
- Court system
- Corrections

Drug and Alcohol Abuse
- Pharmacological principles
- Classification of drugs
- Properties of alcohol, sedative hypnotics, narcotic analgesics, stimulants, hallucinogens and other drugs
- Prevention/treatment

Foundations of Education
- Contemporary issues in education
- Past and current influences on education
- Interrelationships between contemporary issues and influences, past or current, in education

Fundamentals of Counseling
- Counselor roles and functions
- Counseling relationships
- Theoretical approaches
- Social and cultural foundations

General Anthropology
- Anthropology branches, methodologies, and theoretical perspectives
- Physical anthropology and archeology
- Nature of culture
- Social, economic, political, and religious organization

A History of the Vietnam War
- The first Vietnam War (1945-1954)
- LBJ goes to war (1964-1965)
- Tet Offensive (1968)
- Vietnamizing the war (1968-1973)
- Peace is at hand (1968-1973)

Human/Cultural Geography
- Basic facts and concepts
- Culture and environment
- Spatial processes
- Regional geography

Introduction to Law Enforcement
- U.S. criminal justice and police systems
- Issues and organization of police
- Constitutional law and precedents

Introduction to the Modern Middle East
- Geography and early history
- Nineteenth and early twentieth century
- First and Second World wars
- Soviet-U.S. rivalry, Suez Crisis, Iranian Revolution, Persian Gulf Crisis

Lifespan Developmental Psychology
- Biological development
- Perception, learning, and memory
- Social, emotional, and personality development
- Cognition and language

BUSINESS

Business Law II
- Sales of goods
- Debtor and creditor relations
- Business organizations
- Property
- Commercial paper

Business Mathematics
- Basic computation — fractions, percents, ratios, graphs
- Business applications

Introduction to Business
- Economic, international, government, and ownership issues
- Management process
- Human resources and production
- Marketing, finance

Introduction to Computers with Programming in BASIC
- Hardware, software, data management
- Communications and connectivity
- BASIC: I/O, expressions, loops, conditionals, strings, arrays

Management Information Systems
- Systems theory, analysis and design of systems
- Hardware and software
- Database management
- Telecommunications
- Management of the MIS functional area, informational support

Money and Banking
- Role of commercial banks and other intermediaries
- Federal Reserve System
- Macroeconomic activity, U.S. policy, international system

Personal Finance
- Financial goals, budgeting, cash management, economic terminology
- Credit, debt, and major purchases
- Taxes and insurance
- Investments, retirement and estate planning

Personnel/Human Resource Management
- Employment actions
- Training and development
- Performance appraisals
- Compensation and personnel legislation
- Labor relations

Principles of Finance
- Financial statements and ratio analysis
- Capital budgeting
- Time value of money
- Break-even and leverage, common stock

Principles of Financial Accounting
- General concepts and principles
- Financial statements
- Accruals and deferrals
- Current accounts

Principles of Supervision
- Roles and responsibilities
- Management functions
- Legal issues, stress management, unions, productivity and quality concerns

Organizational Behavior
- Individual and group processes and characteristics
- Organizational processes and behavior
- Change and development processes

APPLIED TECHNOLOGY

Technical Writing
- Elements of technical reports
- Style, grammar, mechanics, organizing
- Types of technical reports

HUMANITIES

Ethics in America
- Ethical traditions
- Analysis of issues in relationships
- Tradition versus analysis of issues

Introduction to World Religions
- Dimensions and approaches
- Primal religions, Hinduism, Buddhism, Confucianism, Taoism, Judaism, Christianity, Islam

Principles of Public Speaking
- Topics and purposes
- Organization and supporting materials
- Language, style, and delivery
- Audience analysis and adaptation

PHYSICAL SCIENCE

Astronomy
- The solar system
- The sun and stars
- Our galaxy and the universe
- Measurement

Environment and Humanity: The Race to Save the Planet
- Ecological concepts
- Human impact on the environment
- Environmental management
- Political processes and the future

Here's to Your Health
- Mental health and behavior
- Human development
- Substance use and abuse
- Fitness and nutrition
- Risk factors and disease
- Consumer awareness, environmental concerns

Physical Geology
- Earth materials
- Surface processes
- Internal earth processes
- Applications

Principles of Physical Science I
- Physics: Newton's laws, energy, thermodynamics optics, electricity, and magnetism
- Chemistry: matter, atomic theory and structure, chemical reactions

NOTE:

MOST SCHOOLS DO NOT ACCEPT ALL TESTS FOR ALL PROGRAMS. CHECK WITH THE SCHOOL FROM WHICH YOU EXPECT TO RECEIVE CREDIT.

DANTES

CONTENTS

from the official announcement for instructional purposes

DANTES Subject Standardized Tests (DSSTs) Receive College Credit for Today's Educational Alternatives

Today, there are many educational alternatives to the classroom—you can learn from your job, your reading, your independent study, and special interests you pursue. You may already have learned the materials covered by some college-level courses.

The DANTES Subject Standardized Tests (DSSTs) is a nationally recognized testing program that gives you the opportunity to receive college credit for learning acquired outside the traditional college classroom. The program, developed by The Chauncey Group International, a subsidiary of Educational Testing Service (ETS), is administered year-round by colleges and universities throughout the United States. Thousands of individuals, particularly adults who are interested in boosting their careers, take the DSSTs annually and receive college credit. MAKE IT A POINT to take advantage of the DSST testing program; it speeds the educational process and provides the flexibility adults need, making earning a degree more feasible.

No prerequisites are required to take the DSSTs; however, because requirements differ from college to college, please check with the school where you are enrolled before taking any DSST. Over 1,200 colleges and universities currently award credit for DSSTs, and the number is growing every day. You can choose from over 35 test titles in the areas of Social Science, Business, Mathematics, Applied Technology, Humanities, and Physical Science. A brief description of each examination is found on pages 6 through 8 of this booklet.

Reach Your Career Goals Through Education

Use DSSTs to help you earn your degree, get a promotion, or simply demonstrate that you have college-level knowledge in subjects relevant to your work.

Save Time...

You don't have to sit through classes when you already know through your experiences most of what is being taught and can learn the rest yourself. You might be able to bypass introductory-level courses in subject areas you already know.

Save Money...

DSSTs save you money because the classes you bypass by earning credit through the credit-by-examination program are classes you won't have to pay for on your way to earning your degree. You can use the money instead to take more advanced courses that can be more challenging and rewarding.

Improve Your Chances for College Admission

Each college has its own admission policies; however, having passing scores for DSSTs on your transcript can provide strong evidence of how well you can perform at the college level.

Gain Confidence Performing at a College Level

Many adults returning to college find that lack of confidence is often the greatest hurdle to overcome. DSSTs can show you how successfully you can compete in the college environment because your performance is compared with the performance of college students.

Make Up for Courses You Might Have Missed

You may be ready to graduate from college and find that you are a few credits short of earning your degree. By using semester breaks, vacation time, or leisure time to study independently, you can prepare to take one or more DSSTs and graduate on time.

If You Cannot Attend Regularly Scheduled Classes...

If your lifestyle or responsibilities prevent you from attending regularly scheduled classes, you can earn your college degree from a college offering an external degree program. This type of program allows you to earn your degree by study and experience outside the traditional classroom. Credit is earned by passing credit-by-examinations such as the DSSTs, portfolio assessment, on-line classes, and other distance learning. To learn more, write or call:

Charter Oak College
66 Cedar Street
Newington, Connecticut 01611
(203) 666-4595

Regents College
7 Columbia Circle
Albany, New York 12203-5159
(518) 464-8500

Thomas Edison State College
101 West State Street
Trenton, New Jersey 08608
(609) 984-1150

Other colleges and universities offer external degree or distance learning programs. For additional information, contact the college you plan to attend or:

Center for Adult Learning and Education Credentials
American Council on Education
1 DuPont Circle
Washington, DC 20036
(202) 939-9475

Where to Take DSSTs

DANTES Subject Standardized Tests (DSSTs) are administered at more than 600 locations nationwide. Each location determines the frequency and scheduling of test administrations.

Your college or one near you may be a DSSTs test site. The free booklet DSST Participating Colleges and Universities lists all DSST test sites and their addresses and phone numbers so you can contact them about scheduling testing. In addition to listing DSST test sites, DSST Participating Colleges and Universities also lists colleges that award DSST credits.

DSST Participating Colleges and Universities is available by writing: DANTES Program, The Chauncey Group International, 664 Rosedale Road, Princeton, New Jersey 08540. You may also order by e-mail at DANTES@Chauncey.com. Please be sure to include your complete name and mailing address.

Earning Credit for DSST Examinations

Credit is awarded for passing scores on DSST exams at more than 1,200 colleges and universities in the United States.

To find out whether or not the college of your choice awards credit for passing DSST scores, contact the admissions office or counseling and testing office. The college can also provide information on the scores required for awarding credit, the number of credit hours awarded, and any courses that can be bypassed with satisfactory scores.

It is important that you contact the college of your choice as early as possible because colleges' credit-awarding policies differ.

Arranging for DSST test administration

After you have selected a college that administers DSSTs, you will need to contact their representative listed in the DSST Participating Colleges and Universities to make arrangements for date and time of testing. Please be aware that certain colleges provide DSST testing services to enrolled students only.

The fee to take a DSST is $27 per test. This fee entitles you to two score reports after the test is scored. One will be sent directly to you and the other will be sent directly to the college or university of your choice. You may pay the fee with a certified check or money order made payable to DANTES Program or you may charge the fee to your Visa, MasterCard, or American Express credit card.

In addition, the test site may also require a test administration fee for each examination, to be paid directly to the institution. Contact the test site to determine its fee and payment policy.

Other Testing Arrangements

If you are unable to find a test site near your location, you may want to contact the testing office of a local accredited college or university to determine whether a representative from that office will agree to administer the test(s) to you on a one-time only basis. The school's representative should then contact the DANTES Client Services Representative at (609) 720-6532 to arrange for this special permission.

If you are a civilian at or near a military base and you do not have access to a civilian test center, you may want to contact the base education officer to see whether security and space permit you to test on the base. This decision is based on the discretion of the base representative.

Testing Accommodations for Students with Disabilities

The Chauncey Group International (Chauncey Group) is committed to serving test takers with disabilities by providing services and reasonable testing accommodations as set forth in the provisions of the Americans with Disabilities Act (ADA). If you have a disability, as prescribed by the ADA, and require special testing services or arrangements, please contact the test administrator at the test site. You will be asked to submit to the test administrator documentation of your disability and your request for special accommodations. The test administrator will then forward your documentation along with your request for testing accommodations to the Chauncey Group for approval. Please submit your request as well in advance of your test date as possible so that we may have the opportunity to make the necessary accommodations.

Only test takers with documented disabilities are eligible for special accommodations.

Description of the DSST Examinations

Mathematics

- **Fundamentals of College Algebra** covers mathematical concepts such as fundamental algebraic operations; linear, absolute value; quadratic equations, inequalities, radials, exponents and logarithms, factoring polynomials and graphing.
- **Principles of Statistics** tests the understanding of the various topics of statistics, both qualitatively and quantitatively, and the ability to apply statistical methods to solve a variety of problems. The topics included in this test are descriptive statistics; correlation and regression; and probability.

Social Science

- **Art of the Western World** deals with the history of art during the following periods: classical; Romanesque and Gothic; early Renaissance; high Renaissance, Baroque; rococo; neoclassicism and romanticism; realism, impressionism and post-impressionism; early twentieth century; and post-World War II.
- **Contemporary Western Europe: 1946 - 1990** tests the knowledge of basic facts and terms and the understanding of concepts and principles related to the areas of the historical background of the aftermath of the Second World War and rebuilding of Europe, national political systems, and issues in Western European societies.
- **An Introduction to the Modern Middle East** emphasizes core knowledge (including geography, Judaism, Christianity, Islam, ethnicity); nineteenth-century European impact; twentieth-century Western influences; World Wars I and II; new nations and Soviet-United States rivalry; the Suez Crisis; the Iranian Revolution; the Soviet invasion of Afghanistan, and the Persian Gulf Crisis.
- **Human/Cultural Geography** includes the Earth and basic facts (coordinate systems, maps, physiography, atmosphere, soils and vegetation, water); culture and environment, spatial processes (social processes, modern economic systems, settlement patterns, political geography); and regional geography.
- **Rise and Fall of the Soviet Union** covers Russia under the Old Regime; the Revolutionary Period; New Economic Policy; Pre-war Stalinism; The Second World War; Post-war Stalinism; The Krushchev Years; The Brezhnev Era; and reform and collapse.
- **A History of the Vietnam War** covers the history of the roots of the Vietnam War; the First Vietnam War (1946-1954); pre-war developments (1954-1963); American involvement in the Vietnam War; Tet (1968); Vietnamizing the War (1968-1973); Cambodia and Laos; peace; legacies and lessons.
- **The Civil War and Reconstruction** covers the Civil War from pre-secession (1861) through Reconstruction. It includes causes of the war; secession; Fort Sumter; the war in the east and in the west; major battles; the political situation; assassination of Lincoln; end of the Confederacy; and Reconstruction.
- **Foundations of Education** includes topics such as contemporary issues in education; past and current influences on education (philosophies, democratic ideals, social/economic influences); and the interrelationships between contemporary issues and influences.
- **Life-span Developmental Psychology** covers models and theories; methods of study; ethical issues; biological development; perception, learning and memory; cognition and language; social, emotional, and personality development; social behaviors, family life cycle, extrafamilial settings; singlehood and cohabitation; occupational development and retirement; adjustment to life stresses; and bereavement and loss.
- **Drug and Alcohol Abuse** includes such topics as drug use in society; classification of drugs; pharmacological principles; alcohol (types, effects of, alcoholism); general principles and use of sedative hypnotics, narcotic analgesics, stimulants, and hallucinogens; other drugs (inhalants, steroids); and prevention/treatment.
- **General Anthropology** deals with anthropology as a discipline; theoretical perspectives; physical anthropology; archaeology; social organization; economic organization; political organization; religion; and modernization and application of anthropology.
- **Introduction to Law Enforcement** includes topics such as history and professional movement of law enforcement; overview of the U.S. criminal justice system; police systems in the U.S.; police organization, management, and issues; and U.S. law and precedents.

- **Criminal Justice** deals with criminal behavior (crime in the U.S., theories of crime, types of crime); the criminal justice system (historical origins, legal foundations, due process); police; the court system (history and organization, adult court system, juvenile court, pre-trial and post-trial processes); and corrections.
- **Fundamentals of Counseling** covers historical development (significant influences and people); counselor roles and functions; the counseling relationship; and theoretical approaches to counseling.

Business

- **Principles of Finance** deals with financial statements and planning; time value of money; working capital management; valuation and characteristics; capital budgeting; cost of capital; risk and return; and international financial management.
- **Principles of Financial Accounting** includes topics such as general concepts and principles, accounting cycle and classification; transaction analysis; accruals and deferrals; cash and internal control; current accounts; long- and short-term liabilities; capital stock; and financial statements.
- **Personnel/Human Resource Management** covers general employment issues; job analysis; training and development; performance appraisals; compensation issues; security issues; personnel legislation and regulation; labor relations and current issues.
- **Organizational Behavior** deals with the study of organizational behavior (scientific approaches, research designs, data collection methods); individual processes and characteristics; interpersonal and group processes and characteristics; organizational processes and characteristics; and change and development processes.
- **Principles of Supervision** deals with the roles and responsibilities of the supervisor; management functions (planning, organization and staffing, directing at the supervisory level); and other topics (legal issues, stress management, union environments, quality concerns).
- **Business Law II** covers topics such as sales of goods; debtor and creditor relations; business organizations; property; and commercial paper.
- **Introduction to Computers with Programming** in Basic includes topics such as history and technological generations; hardware/software; applications to information technology; program development; data management; communications and connectivity; computing and society; and programming in BASIC.
- **Management Information Systems** covers systems theory, analysis and design of systems, hardware and software; database management; telecommunications; management of the MIS functional area and informational support.
- **Introduction to Business** deals with economic issues affecting business; international business; government and business; forms of business ownership; small business, entrepreneurship and franchise; management process; human resource management; production and operations; marketing management; financial management; risk management and insurance; and management and information systems.
- **Money and Banking** covers the role and kinds of money; commercial banks and other financial intermediaries; central banking and the Federal Reserve system; money and macroeconomics activity; monetary policy in the U.S.; and the international monetary system.
- **Personal Finance** includes topics such as financial goals and values; budgeting; credit and debt; major purchases; taxes; insurance; investments; and retirement and estate planning.
- **Business Mathematics** deals with basic operations with integers, fractions, and decimals; round numbers; ratios; averages; business graphs; simple interest; compound interest and annuities; net pay and deductions; discounts and markups; depreciation and net worth; corporate securities; distribution of ownership; and stock and asset turnover.

Physical Science

- **Astronomy** covers the history of astronomy, celestial mechanics; celestial systems; astronomical instruments; the solar system; nature and evolution; the galaxy; the universe; determining astronomical distances; and life in the universe.
- **Here's to Your Health** covers mental health and behavior; human development and relationships; substance abuse; fitness and nutrition; risk factors, disease, and disease prevention; and safety, consumer awareness, and environmental concerns.

- **Environment and Humanity** deals with topics such as ecological concepts (ecosystems, global ecology, food chains and webs); environmental impacts; environmental management and conservation; and political processes and the future.

- **Principles of Physical Science I** includes physics: Newton's Laws of Motion; energy and momentum; thermodynamics; wave and optics; electricity and magnetism; chemistry: properties of matter; atomic theory and structure; and chemical reactions.

- **Physical Geology** covers Earth materials; igneous, sedimentary, and metamorphic rocks; surface processes (weathering, groundwater, glaciers, oceanic systems, deserts and winds, hydrologic cycle); internal Earth processes; and applications (mineral and energy resources, environmental geology).

Applied Technology

- **Technical Writing** covers topics such as theory and practice of technical writing; purpose, content, and organizational patterns of common types of technical documents; elements of various technical reports; and technical editing. Students may write a short essay on one of the technical topics provided. The essay will not be scored by the Chauncey Group; however, a copy of the essay will be provided with the score report or transcript.

Humanities

- **Ethics in America** deals with ethical traditions (Greek views, Biblical traditions, moral law, consequential ethics, feminist ethics); ethical analysis of issues arising in interpersonal and personal-societal relationships and in professional and occupational roles; and relationships between ethical traditions and the ethical analysis of situations. Students may write an essay to analyze a morally problematic situation in terms of issues relevant to a decision and arguments for alternative positions. The essay will not be scored by the The Chauncey Group; however, a copy of the essay will be provided with the score report or transcript.

- **Introduction to World Religions** covers topics such as dimensions and approaches to religion; primal religions; Hinduism; Buddhism; Confucianism; Taoism; Judaism; Christianity; and Islam.

- **Principles of Public Speaking** consists of two parts: Part One requires the student to record an impromptu persuasive speech that will be scored. Part Two consists of multiple-choice questions covering considerations of public speaking; audience analysis; purposes of speeches; structure/organization; content/supporting materials; research; language and style; delivery; communication apprehension; listening and feedback; and criticism and evaluation.

HOW TO ORDER FACT SHEETS/STUDY GUIDES

For each test, there is a Fact Sheet/Study Guide that gives information about the topics that will be covered, books that would be useful for review, the number of questions on the test, and the passing score recommended by the American Council on Education (ACE) for granting credit. It is recommended that students check with the school where they are enrolled about the school's requirements for earning credit. Some schools set higher standards than the ACE minimum recommendation. The Fact Sheet/Study Guides also contain sample test questions to give you an idea of what the test is like. You may order a free copy of the Fact Sheet/Study Guide you need by completing and mailing in the Fact Sheet/Study Guide Order Form on page 11 or sending an e-mail request to DANTES@Chauncey.com. When sending an e-mail request, please include your name and complete mailing address.

ON THE DAY OF THE EXAMINATION

Here are a few tips to help you on the day of your examination:

- Arrive on time as a courtesy to the test administrator. Remember, many test administrators have responsibilities outside the testing office and may have a specific amount of time allotted for your test administration.

- Have your personal photo identification with you. <u>Anyone who fails to present a photo I.D. will not be tested.</u>

- Bring several No.2 (soft-lead) pencils with good erasers, a watch, and a black pen if you will be writing an essay.

- Do not bring books or papers.

- Do not bring an alarm watch that beeps, a telephone, or a phone beeper into the testing room.

- The use of non-programmable calculators, slide rules, scratch paper and/or other materials is permitted for some of the tests. Please refer to the section entitled Auxiliary Test Materials on page 15 for specific information.

DANTES SUBJECT STANDARDIZED TESTS SCORING POLICIES

Your DSST examination scores are reported only to you, unless you request that they be sent elsewhere. If you want your scores sent to a college you must grid the correct code number or write the name and address of the school on your answer sheet at the time you take the test.

Receiving Your Score Report

DSST answer sheets are scanned by computer. Allow four weeks after testing to receive your score report.

Do not call before the allotted time for scoring has elapsed—this will not expedite the processing of your scores. Please note that scores will not be reported to students over the telephone because of privacy and security considerations.

Scoring of Public Speaking Speech and Retention of Essays

The speech portion of Public Speaking will be sent to speech raters who are faculty members at accredited colleges that currently teach the course. These raters have received special training to ensure uniformity. Because of this, scores for the Public Speaking examination are normally available in 6 - 8 weeks. If you take the Public Speaking examination and fail, you may retest after 6 months (180 days) on a different speech topic. You must wait 2 years to re-test using the same speech topic.

The essays for Ethics in America and Technical Writing are optional and thus are not scored by raters. The essays are forwarded to the colleges and universities that you designate, along with your score report, for their use in determining the award of credit.

How to Get Transcripts

To request a transcript, use the form located on page 13. There is a $10 fee for each transcript you request.

Availability/Use of Scores

College score requirements for awarding credit vary. Please contact the college or university in which you are enrolled for information on the minimum score it requires to grant credit.

Candidate scores that have been released to colleges may be used for educational research or validity studies by the receiving colleges or by The Chauncey Group International, but the names of individuals will not be revealed or attached to particular scores.

Score Validity

The DANTES Program is obligated to report scores that accurately reflect the performance of the test taker. The program maintains test administration and test security standards designed to assure that all test takers are given the same opportunity to demonstrate their abilities and to prevent any test taker from gaining an unfair advantage.

The DANTES Program reserves the right to cancel any test score if the test taker engages in misconduct, if there is a testing irregularity, or if there is reason to question the score's validity.

Reviews of questionable scores are confidential. If it is necessary to cancel scores that have already been reported, score users are notified, but the reasons for cancellation are not disclosed.

FREQUENTLY ASKED QUESTIONS

How should I study for the examinations?

For each of the DSSTs there is a Fact Sheet/Study Guide which outlines the topics covered by each test. Each Fact Sheet/Study Guide includes a list of sample questions and a list of recommended references for you to use. You may order a free copy of the Fact Sheet/Study Guides you need by completing and mailing the Fact Sheet/Study Guide Order Form on page 11 or by sending an e-mail request to DANTES@chauncey.com. Please include your name and complete mailing address.

In order to pass the test, must I study from one of the recommended references?

The recommended references are a listing of books that were currently being used as text books for the same course title at the time the test was written. If some time has passed since then, you may have difficulty finding the exact title or edition of the book. You might want to consider visiting the bookstore of a local college or university and comparing the text book they currently use for the same class title to the topics listed on the Fact Sheet/Study Guide. If the topics listed on the Fact Sheet/Study Guide are covered by the selected textbook, then you should be reasonably prepared for the test by studying that text book.

Is there a penalty for guessing on the tests?

There is no penalty for guessing on the DANTES Subject Standardized Tests (DSSTs), so you should mark an answer for each question.

How much time will I have to complete the test?

Many DSSTs can be completed within 90 minutes; however, additional time can be allowed if necessary.

What should I do if I find an error in a test question?

If you think there is an error in a test question or any aspect of a test administration, continue with the test, but report the information to the testing supervisor. This may be done after the test by asking that it be noted on the Supervisor's Irregularity Report or by writing to DANTES Program, The Chauncey Group International, Princeton, NJ 08540 and indicating the form and question number(s) or circumstances as well as your name and address.

Will my test scores be released without my permission?

Your test score will not be released to anyone other than the school you designate on your answer sheet unless you write to us and ask us to send a transcript elsewhere. Instructions about how to do this can be found on your score report. Your scores may be used for research purposes, but individual scores are never made public nor are individuals identified if research findings for a group are made public.

If I do not achieve a passing score on the test, how long must I wait until I can take the test again?

If you do not receive a score on the test that will enable you to obtain credit for the course, you may take the test again after six months (180 days). Please do not attempt to take the test before six months (180 days) have passed because you will receive a score report marked invalid and will have wasted the time and the money you invested in retesting. There must be 180 days between test dates in order for your score report to be valid.

Can my test scores be canceled?

The test administrator is required to report any irregularities to the DANTES Program. The consequence of bringing unauthorized materials into the testing room or giving or receiving help will be the forfeiture of your test fee. The DANTES Program reserves the right to cancel scores and not issue score reports in such situations.

What can I do if I feel that my test scores were not accurately reported?

If you have reason to believe that your score(s) were not accurately reported, you may request to have your answer sheet reviewed and hand scored. There is a $10 fee for this service. Please submit a check or money order for $10 payable to DANTES Program with your request in writing for score verification to: DANTES Program, P.O. Box 6604, Princeton, NJ 08541-6604. Include your full name, the test title, the date you took the test, and your Social Security number. We will notify you of the results of hand scoring and, if your scores were not accurate, a corrected score report will be forwarded to you and to any school to which score reports were sent. In the event that a scoring error occurred, your $10 fee will be refunded.

What does ACE recommendation mean?

The ACE recommendation is the minimum passing score recommended by the American Council on Education for any given test. It is equivalent to the average score of students in the DANTES norming sample who received a grade of C for the course. Some schools require a score higher than the ACE recommendation.

How long are my test scores kept on file?

Test scores are kept on file for a period of 20 years.

REMEMBER TO MAKE IT A POINT

You are always learning, whether it be inside a classroom, on your job, or running a household. You can gain experience through retail sales experience, parenting, reading, hobbies (such as Toastmaster's), and non-credit business training. MAKE IT A POINT to get credit for what you already know with DANTES Subject Standardized Tests.

DANTES SUBJECT STANDARDIZED TESTS (DSSTs)

FACT SHEET ❖ ORDER FORM

INSTRUCTIONS:

Please ship the quantities of the DANTES Fact Sheets on the reverse side to:

Name ____

Address ____

City ______________________ **State** ______ **Zip Code** __________

Today's Date ____________________

To place your order, please mail the completed form to:

The DANTES Program
The Chauncey Group International
PO Box 6604, MS-23P
Princeton, NJ 08541-6604

Fax: (609) 720-6800
For confirmation of receipt of FAX, call: 609-720-6726

For information on The DANTES Program,
call (609) 720-6740.

Fact Sheets for DANTES Subject Standardized Tests

Quantity	Title
	MATHEMATICS
______	Fundamentals of College Algebra
______	Principles of Statistics
	SOCIAL SCIENCE
______	Art of the Western World
______	Contemporary Western Europe: 1946-1990
______	An Introduction to the Modern Middle East
______	Human/Cultural Geography
______	Rise and Fall of the Soviet Union
______	A History of the Vietnam War
______	The Civil War and Reconstruction
______	Foundations of Education
______	Lifespan Developmental Psychology
______	General Anthropology
______	Drug and Alcohol Abuse
______	Introduction to Law Enforcement
______	Criminal Justice
______	Fundamentals of Counseling
	BUSINESS
______	Principles of Finance
______	Principles of Financial Accounting
______	Personnel/Human Resource Management
______	Organizational Behavior
______	Principles of Supervision
______	Business Law II
______	Introduction to Computers with Programming in BASIC
______	Introduction to Business
______	Money and Banking
______	Personal Finance
______	Management Information Systems
______	Business Mathematics

Quantity	Title
	PHYSICAL SCIENCE
______	Astronomy
______	Here's to Your Health
______	Environment and Humanity: The Race to Save the Planet
______	Principles of Physical Science I
______	Physical Geology
	APPLIED TECHNOLOGY
______	Technical Writing
	HUMANITIES
______	Ethics in America
______	Introduction to World Religions
______	Principles of Public Speaking

DANTES SUBJECT STANDARDIZED TESTS (DSSTs)

CIVILIAN TRANSCRIPT REQUEST FORM FOR DSST SCORES

TRANSCRIPT REQUEST FORM FOR DSST SCORES

To avoid a delay, do not send a letter with this form.

FEE:

A **$10** fee is charged for **each** transcript requested. A transcript may include any or all DSST scores.

Enclose a check or money order payable to DANTES Program. If you prefer to pay for your transcript with a credit card, please supply the information below. Your account will be charged with the total amount due. Incomplete forms or forms received without fees will be returned to the requester. **DO NOT SEND CASH OR STAMPS.**

SEND COMPLETED FORM TO:

DANTES Program
The Chauncey Group International
P.O. Box 6604
Princeton, NJ 08541-6604

☐ Check or Money Order enclosed (In U.S. Dollars) Please charge my ☐ MasterCard ☐ Visa ☐ American Express

My credit card number is: Expiration date: ____ / ____ (Month / Year)

___ ___ ___ ___ - ___ ___ ___ ___ - ___ ___ ___ ___ - ___ ___ ___ ___

PLEASE PRINT ALL INFORMATION REQUESTED BELOW.

Last Name (Include Maiden Name or Former Last Names) First MI / Social Security Number

Present Address Number and Street / Date of Birth

City and State Zip Code Phone Number

I understand that my signature permits DANTES to release my transcript to the designated recipient(s).

(Signature)

Please prepare my transcript and include the following (check only one):

☐ Scores on all tests ☐ Only test scores that are at or above the ACE Recommended Minimum Score ☐ Only scores on tests named below

TEST TITLES:

Approximate Date of Last DANTES Test

☐ Please send a copy of my transcript to the following school(s) (**$10.00 per transcript**):

School Name	School Name
Attn:	Attn:
Address	Address
City & State Zip Code	City & State Zip Code

Notes for Test Takers

Here is some information about the DANTES test you are planning to take that you may find helpful.

Preparing for the Test

For each test, there is a fact sheet that gives information about the topics that will be covered, books that would be useful for review, the number of questions on the test, and the minimum score recommended by the American Council on Education (ACE) for credit. (You should check the passing score with your school; some have set higher standards than the ACE minimum recommendation.) Some of the fact sheets also contain sample test questions to give you an idea of what the test is like. Copies of these sheets are available without charge from the test administrator, your counselor, by phone at 609-720-6740 or by e-mail at DANTES@chauncey.com.

Guessing

There is no penalty for guessing on the DANTES Subject Standardized Tests, so you should mark an answer for each question.

Test Timing

These tests are untimed. It generally takes about 1 1/2 hours to take a DANTES Subject Standardized Test; however, you may work longer if necessary.

Test Materials

The use of non-programmable calculators, slide rules, scratch paper and/or other materials is permitted for some of the tests. The test administrator has information about whether the test(s) you are planning to take fall into this category and which materials are permitted. If you wish to bring any materials into the test room, please check before testing to be sure they are permitted.

DANTES

Subject Standardized Tests

Test Administration

If you think there is an error in a test question, continue with the test, but report the information to The Chauncey Group International. This may be done by telling the supervisor after the test, asking that it be noted on the Supervisor's Irregularity Report, by writing to **The Chauncey Group, DANTES Program, CN-6604, Princeton, NJ 08541-6604**, indicating the form and question numbers as well as your name and address.

Confidentiality

Your test score on your test will not be released to anyone other than the school you designate on your answer sheet unless you write to us and ask us to send a transcript elsewhere. Instructions about how to do this will be found on your score report. Your scores may be used for research purposes, but individual scores are never made public nor are individuals identified if research findings for a group are made public.

Students with Disabilities

Special testing arrangements are authorized for students with disabilities. If you are in need of this service, speak to your test administrator in advance of the testing. A reader or writer may be used or other special arrangements may be made to minimize the effect of your disability on your score.

Retesting

If you do not receive a score on the test that will enable you to obtain credit for the course, you may take the test again after six months (180 days). Please do not attempt to retake the test before six months (180 days) have elapsed because you will receive a score report marked invalid and will have wasted the time and money you invested in retesting.

Score Cancellation

The test administrator is required to report any irregularities to The Chauncey Group. The consequences of bringing unauthorized materials into the testing room or giving or receiving help will be the forfeiture of your test fee. The Chauncey Group reserves the right to cancel scores and not issue score reports in such situations.

Score Verification

If you have reason to believe that your score(s) were not accurately reported, you may request to have your answer sheet hand scored. There is a $10 fee for this service. Please submit a check or money order for $10 payable to **DANTES Program** with your request in writing for score verification to: **The Chauncey Group International, P.O. Box 6604, Princeton, NJ 08541-6604**. Include your full name, the test title, the date you took the test and your Social Security number. We will notify you of the results of hand scoring and, if your scores were not accurate, a corrected score report will be forwarded to you and any school, to which score reports were sent. In the event that a Chauncey Group error occurred, your money will be refunded.

DANTES
Subject Standardized Tests

ACE Credit Recommendations

American Council on Education Credit Recommendations

Test Form Number	Title	Credit Amount in Semester Hours	Minimum Score
MATHEMATICS			
SF/SG 424	Fundamentals of College Algebra	3B	47
SE 450	Principles of Statistics	3B	48
SOCIAL SCIENCE			
SE 461	Art of the Western World	3B	48
SE 465	Contemporary Western Europe: 1946-1990	3B	48
SE 469	An Introduction to the Modern Middle East	3B	44
SF/SG 470	Human/Cultural Geography	3B	48
SE 471	Rise and Fall of the Soviet Union	3BU	45
SF 473	A History of the Vietnam War	3B	49
SE/SF 483	The Civil War and Reconstruction	3BU	47
SE 489	Foundations of Education	3B	46
SF/SG 490	Lifespan Developmental Psychology	3B	46
SF 494	General Anthropology	3B	47
SE 495	Drug and Alcohol Abuse	3BU	49
SG/SH 497	Introduction to Law Enforcement	3B	45
SF/SG 498	Criminal Justice	3B	49
SG/SH 562	Fundamentals of Counseling	3B	45
BUSINESS			
SF/SG 524	Principles of Finance	3BU	46
SF 525	Principles of Financial Accounting	3B	49
SF 530	Personnel/Human Resource Management	3B	48
SF/SG 531	Organizational Behavior	3B	48
SE/SF 532	Principles of Supervision	3B	46
SE 534	Business Law II	3BU	52
SG/SH 536	Introduction to Computing	3B	45
SE/SF 543	Introduction to Business	3B	46
SG/SH 548	Money and Banking	3BU	48
SE 550	Personal Finance	3B	46
SE 551	Management Information Systems	3BU	46
SF/SG 812	Business Mathematics	3B	48

Test Form Number	Title	Credit Amount in Semester Hours	Minimum Score
PHYSICAL SCIENCE			
SF 500	Astronomy	3B	48
SF/SG 508	Here's to Your Health	3B	48
SF/SG 511	Environment and Humanity: The Race to Save the Planet	3B	46
SE 512	Principles of Physical Science I	3B	47
SF 519	Physical Geology	3B	46
APPLIED TECHNOLOGY			
SF/SG 820	Technical Writing *(TWE Technical Writing Essay)*	3B	46
HUMANITIES			
SF/SG 474	Ethics in America *(SF 474 Ethics in America Essay)*	3B	46
SE 496	Introduction to World Religions	3B	49
SE/SF 815*	Principles of Public Speaking	3B	47

*In addition to a minimum score of 47 on the multiple-choice test, an examinee must also receive a passing grade on the speech.

B = Baccalaureate Program
BU = Baccalaureate Upper Division

ABOUT THE DANTES PROGRAM

The DANTES Subject Standardized Tests (DSSTs) enable students to obtain college credit for what they have learned on the job, through self-study, personal interest, correspondence courses or by any other means. The DANTES program is available through The Chauncey Group International. It is used by colleges and universities to award college credit to students who demonstrate that they know as much as students completing an equivalent college course. It is a cost-efficient, time-saving way for students to use their knowledge to accomplish their educational goals.

DANTES COLLEGES

This booklet lists the names and addresses of colleges and universities that award credit for passing scores on DSSTs. Those that both administer tests and accept scores are identified with an asterisk. Please note:

- Most schools accept the American Council on Education (ACE) recommendations for the minimum score required and the amount of credit awarded; however, not all schools do. Be sure to check your schools policy regarding the score required for credit and the number of credits to be awarded.
- Not all tests are accepted by all institutions. Even when a test is accepted by an institution, it may not be acceptable for every program at that institution. Before considering testing, ascertain the acceptability of a specific test for your particular program.
- Colleges and universities that administer DSSTs may administer them to any applicant—or they may administer the tests only to students registered at their institution. Decisions about who will be allowed to test are made by the school. Students should contact the test center to determine current policies and schedules for DSST testing.

PREPARING FOR THE TEST

There are approximately 35 DANTES Subject Standardized Tests covering a wide range of college-level academic, business, and technical subjects. For each test, there is a Fact Sheet/Study Guide containing a description of the test content, sample questions, a list of texts appropriate for review, and technical information regarding how the passing score was established. Fact Sheets are available free of charge from test administrators, or directly from The Chauncey Group International; write to: DANTES Program, The Chauncey Group International, 664 Rosedale Road, Princeton, New Jersey 08540, or e-mail: DANTES@Chauncey.com to order them.

TESTING ARRANGEMENTS

Colleges and universities authorized to administer DSSTs usually do so throughout the calendar year. Each school sets its own fee for test administration and establishes its own testing schedule. Contact the representative at the administering school directly to make arrangements for testing. (Their names and telephone numbers are listed in this booklet.)

ALABAMA

9399
ARTHUR BARNETT
DIRECTOR OF ADMISSIONS
ALABAMA STATE UNIVERSITY
915 SOUTH JACKSON ST
MONTGOMERY, AL 36195
205-293-4473

*9048
DR. ROBERT D. SEARCY
COORDINATOR OF TESTING
ATHENS STATE COLLEGE
300 NORTH BEATY ST
ATHENS, AL 35611
256-233-6540

*9246
GAIL BALL
TESTING COORDINATOR
AUBURN UNIVERSITY MONTGOMERY
128 UNIVERSITY CENTER
7300 UNIVERSITY DRIVE
MONTGOMERY, AL 36117
334-244-3758

9093
ALICE GOINS
REGISTRAR
CHATTAHOOCHEE VALLEY STATE COMMUNITY COLLEGE
2602 COLLEGE DR
PHENIX CITY, AL 36867
205-291-4900

8451
JAMES R. INNES
CURRICULUM DIRECTOR
COMMUNITY COLLEGE OF THE AIR FORCE
130 WEST MAXWELL BLVD
MAXWELL AFB, AL 36112-6613
334-953-2876

9144
DAVID CHALKER
REGISTRAR
ENTERPRISE STATE JUNIOR COLLEGE
PO BOX 1300
ENTERPRISE, AL 36331
205-347-2623

*8036
DEBBIE WATKINS
DIRECTOR, ASSESSMENT CENTER
FAULKNER UNIVERSITY
5345 ATLANTA HIGHWAY
MONTGOMERY, AL 36109
334-260-6247

9157
W. BRYAN STONE
REGISTRAR
GADSDEN STATE COMMUNITY COLLEGE
GEORGE WALLACE DR - PO BOX 227
GADSDEN, AL 35902-0227
256-549-8259

*9794
CHARLES THOMPSON
ABLE PROGRAM OFFICE
INTERNATIONAL BIBLE COLLEGE
PO BOX IBC
FLORENCE, AL 35630
(205) 766-6610

918[illegible]
JEANE REYNOLDS
JOHN C. CALHOUN STATE COMMUNITY COLLEGE
PO BOX 2216
DECATUR, AL 35609
205-306-2596

8323
ROGER McALLISTER, COORD
EXTERNAL DEGREE PROGRAM
NEW COLLEGE-UNIVERSITY OF ALABAMA
PO BOX 870182
TUSCALOOSA, AL 35487
205-348-6010

8647
SHARON WILLIAMS
DIRECTOR OF STUDENT SERVICES
NORTHWEST ALABAMA COMMUNITY COLLEGE
RT 3 BOX 77
PHIL CAMPBELL, AL 35581
205-993-5331

*8321
SHIRLEY SCOTT
COUNSELING AND TESTING
OAKWOOD COLLEGE
5022 OAKWOOD RD
HUNTSVILLE, AL 35896

*9402
SHEILA BENNETT
TEST COORDINATOR
TROY STATE UNIVERSITY
501 UNIVERSITY DR
DOTHAN, AL 36303
205-983-6556

*8627
MARY HINSON COOK
STUDENT AFFAIRS
TROY STATE UNIVERSITY
231 MONTGOMERY ST
MONTGOMERY, AL 36104
334-241-9532

9344
VICKIE MILES
DIRECTOR EDUCATIONAL SERVICES
TROY STATE UNIVERSITY
UNIVERSITY AVE
TROY, AL 36082
205-670-3170

9948
MARYANNE IVEY
OFFICE STUDENT TEST/EVAL
UNIVERSITY OF ALABAMA
924 SOUTH 19th ST
BIRMINGHAM, AL 35294-2100
205-934-5503

*9348
SYBIL A. HOBSON
ASSOCIATE DIRECTOR
UNIVERSITY OF ALABAMA
205 STUDENT SERVICES CENTER
801 CAMPUS DRIVE
TUSCALOOSA, AL 35487-0108
205-348-6760

*8667
DAVID E. POWLEY
DIRECTOR OF TESTING
UNIVERSITY OF MOBILE
BOX 13220
MOBILE, AL 36613
205-675-5990

*8619
ELAINE ROWELL
LEARNING ENHANCEMENT SPEC
UNIVERSITY OF NORTH ALABAMA
OFFICE OF STUDENT LIFE, RM 202
GUILLOT UNIVERSITY CENTER
WESLEYAN AVE - PO BOX 5033
FLORENCE, AL 35632
256-765-4248

9198
CLARENCE EGBERT
REGISTRAR
UNIVERSITY OF WEST ALABAMA
LIVINGSTON, AL 35470
205-652-3587

*9984
DORI MORGAN
TEST ADMINISTRATOR
WALLACE STATE COMMUNITY COLLEGE
801 MAIN ST, NORTHWEST
HANCEVILLE, AL 35077
205-352-8000

ALASKA

8394
MARGARET PYLANT
SITE DIRECTOR
CENTRAL TEXAS COLLEGE
KISKA HALL BLDG 658
FT. RICHARDSON, AK 99577

*8139
DIANE TAYLOR
LEARNING CENTER COORDINATOR
KENAI PENINSULA COLLEGE
34820 COLLEGE DR
SOLDOTNA, AK 99669
907-262-0300

*9664
CHRISTINE JAMIN
STUDENT SERVICES COUNSELOR
KODIAK COLLEGE
117 BENNY BENSON DRIVE
KODIAK, AL 99615
907-486-4161

*8576
DEBRA VANASSE
LEARNING ASSISTANCE CENTER
KUSKOKWIM COMMUNITY COLLEGE
BOX 368
BETHEL, AK 99559
907-543-4500

*8578
DOUGLAS GREEN
COUNSELING AND TESTING
MATANUSKA-SUSITNA COLLEGE
PO BOX 2889
PALMER, AK 99645
907-745-9774

9869
MARY GOFF
REGISTRAR
SHELDON JACKSON COLLEGE
801 LINCOLN ST
SITKA, AK 99835
907-747-5216

*9043
JOHN SVIHRA
ASSESSMENT COORDINATOR
UNIVERSITY OF ALASKA
3211 PROVIDENCE DR
ANCHORAGE, AK 99508
907-786-4500

*9403
NATASHA ZAHN
EDUCATION TECHNICIAN
LEARNING AND TESTING CENTER
UNIVERSITY OF ALASKA
11120 GLACIER HWY
JUNEAU, AK 99801
907-465-6438

*8626
KAREN CARLISLE
TEST CENTER COORDINATOR
UNIVERSITY OF ALASKA-BRISTOL BAY
PO BOX 1070
DILLINGHAM, AK 99576
907-842-5901

9349
PATTY.M. WHITE
TESTING COORDINATOR
UNIVERSITY OF ALASKA
112 SIGNERS' HALL
FAIRBANKS, AK 99775
907-474-5277

*8295
DAWN MONTGOMERY
COORDINATOR & TEST CENTER SUPERVISOR
UNIVERSITY OF ALASKA SOUTHEAST
2600 SEVENTH AVENUE
KETCHIKAN, AK 99901-5798
907-228-4524

*8995
LYNN JOHNSON
DIRECTOR, CHUKCHI CAMPUS
UNIVERSITY OF ALASKA-FAIRBANKS
PO BOX 297
KOTZEBUE, AK 99752
907-442-3400

ARIZONA

*9868
STEVE BEAVER
OFFICE OF ADMISSIONS
AMERICAN GRADUATE SCHOOL OF INTERNATIONAL MANAGEMENT
15249 NORTH 59th AVE
GLENDALE, AZ 85306-6000
602-978-7133

*9559
BARBARA J. STRUBLE
POST SECONDARY EDUCATION COORDINATOR
CENTRAL ARIZONA DETENTION CENTER
1155 NORTH PINAL PARKWAY
P.O. BOX 1048
FLORENCE, AZ 85232
520-868-3668

8173
KATHLEEN ROSE
DIRECTOR
CHAPMAN UNIVERSITY
PO BOX 15132
DAVIS MONTHAN AFB
TUCSON, AZ 85708
520-745-6324

8625
OFFICE OF ADMISSIONS
EMBRY-RIDDLE AERONAUTICAL UNIVERSITY
3200 WILLOW CREEK RD
PRESCOTT, AZ 86301-3720

8492
CATHERINE L. MESCHKE
DIRECTOR OF RECORDS AND REGISTRATION
GATEWAY COMMUNITY COLLEGE
108 NORTH 40th ST
PHOENIX, AZ 85034

9407
MARY BAYLESS
DEAN OF ADMISSIONS
GLENDALE COMMUNITY COLLEGE
6000 WEST OLIVE AVE
GLENDALE, AZ 85302
602-435-3000

9406
FAITH A. WEESE
REGISTRAR
GRAND CANYON UNIVERSITY
3300 WEST CAMELBACK RD
PHOENIX, AZ 85017
602-249-3300

8313
MARTHA PENDLETON
ADVISEMENT COUNSELOR
JOHN BROWN UNIVERSITY
FORT SMITH CENTER
4500 TOWSON AVE, STE 216
FORT SMITH, AR 72901
501-646-0082

9901
GENE PARRISH
DIRECTOR, COUNSELING
MESA COMMUNITY COLLEGE
1833 W SOUTHERN AVE
MESA, AZ 85202
602-461-7428

9256
JANE LEMON
OFFICE SPECIALIST SR.
NORTHERN ARIZONA UNIVERSITY
BOX 4103
FLAGSTAFF, AZ 86011
602-523-2108

*8589
JANET MANTIA
ADMISSIONS
OTTAWA UNIVERSITY
2340 WEST MISSION LANE
PHOENIX, AZ 85021
602-371-1188

8404
WILLIAM J. BARNES
ADMINISTRATOR
PARK COLLEGE
58 MSS/DPE
LUKE AFB, AZ 85309
602-935-4424

*9280
PAMELA GYLLING
ASSESSMENT CENTER SPECIALIST
PIMA COMMUNITY COLLEGE
1255 NORTH STONE AVE
TUCSON, AZ 85709
520-206-6370

9871
ELIZABETH SMITH
REGISTRAR
PRESCOTT COLLEGE
220 GROVE AVE
PRESCOTT, AZ 86303
520-776-5164

*9931
LINDA LUKEY
ACADEMIC ADVISOR
RIO SALADO COMMUNITY COLLEGE
2323 WEST 14th ST
TEMPE, AZ 85281
602-517-8553

9363
MONTEREY SIMS
ASSOC. DIR OF ADMISSIONS
UNIVERSITY OF PHOENIX
4615 EAST ELWOOD ST
PHOENIX, AZ 85072-2069
602-927-0099, X1258

*8371
SUSAN McGIVERN
DIRECTOR OF ADMISSIONS
WESTERN INTERNATIONAL UNIVERSITY
9215 NORTH BLACK CANYON HWY
PHOENIX, AZ 85021
602-943-2311

ARKANSAS

*8587
KAREN PITTMAN
PSYCHOMETRIST
ARKANSAS TECH UNIVERSITY
STUDENT DEVELOPMENT CENTER
113 WILLIAMSON HALL
RUSSELLVILLE, AR 72801
501-968-0382

*8865
RON CHESSER
ADMISSIONS COUNSELOR
GARLAND COUNTY COMMUNITY COLLEGE
1 COLLEGE DR
HOT SPRINGS, AR 71913
501-767-9371

*8066
FRAN CALVERT
ASST. REGISTRAR
JOHN BROWN UNIVERSITY
2000 W UNIVERSITY
SILOAM SPRINGS, AR 72761
501-524-7138

*8313
SHERRI WALKER
COUNSELOR
JOHN BROWN UNIVERSITY
ADVANCE PROGRAM
FT. SMITH CENTER
4500 TOWSON AVE, SUITE 216
FT. SMITH, AR 72901
501-646-0082

*8299
LISA KELLEY
SITE DIRECTOR
JOHN BROWN UNIVERSITY
LITTLE ROCK CENTER
5107 WARDEN ROAD, SUITE 4
NORTH LITTLE ROCK, AR 72116
501-771-9000

9230
MYLAS JEFFERS
DIRECTOR, COUNSELING SERVICES
MISSISSIPPI COUNTY COMMUNITY COLLEGE
PO BOX 1109
BLYTHEVILLE, AR 72316
501-762-1020

8410
SALLY HUNTER
COORDINATOR
PARK COLLEGE
PO BOX 1280
JACKSONVILLE, AR 72078
501-988-5624

8790
SONNY WHITTINGTON
DIRECTOR OF ADMISSIONS
SOUTHERN ARKANSAS UNIVERSITY
SAU BOX 1382
MAGNOLIA, AR 71753
501-235-4040

*8701
DR. YASSAMAN MIRDAMADI
UNIVERSITY OF ARKANSAS
730 HOTZ HALL
FAYETTEVILLE, AR 72701
501-575-3948

8949
KWURLY M. FLOYD
REGISTRAR AND ADMISSIONS
UNIVERSITY OF ARKANSAS
BOX 17 - 1200 NORTH UNIVERSITY
PINE BLUFF, AR 71601
501-541-6500

9387
PEGGY FORSBERG
REGISTRAR
WESTARK COMMUNITY COLLEGE
PO BOX 3649
FORT SMITH, AR 72913
501-788-7234

CALIFORNIA

9039
NORMA RAZO
REGISTRAR
ALLAN HANCOCK COLLEGE
800 SOUTH COLLEGE DR
SANTA MARIA, CA 93454
805-922-6966

8462
HENRY DAVIS
COUNSELING OFFICE
ALLAN HANCOCK COLLEGE
PO BOX 5187
VANDENBERG AFB, CA 93437
805-922-6966

*8039
DAVID LAM
PROGRAM DIRECTOR
AMERICAN COLLEGE OF NURSING
2022 TARAVAL ST, SUITE 3253
SAN FRANCISCO, CA 94116
415-661-5533

*8704
ANDREA RICHARDS
DIR. OF ADMISSIONS
ANTIOCH UNIVERSITY - LOS ANGELES
13274 FIJI WAY
MARINA DEL RAY, CA 90292
310-578-1080, X 206

9053
JOSEPH CLARK
ASST. DEAN OF INSTRUCTION
BARSTOW COMMUNITY COLLEGE
2700 BARSTOW RD
BARSTOW, CA 92311
619-252-2411

*8819
MARIEL DELUCA VOTH
DIRECTOR, PACE
BETHEL COLLEGE
6116 AROSA STREET
SAN DIEGO, CA 92115

9890
PAMELA JOLICOEUR
ADULT DEGREE PROGRAM
CALIFORNIA LUTHERAN UNIVERSITY
60 WEST OLSEN RD
THOUSAND OAKS, CA 91360
805-492-2411

*8267
WILLIAM A. JAYO, INSTRUCTOR
EDUCATION DEPARTMENT
CALIFORNIA MEDICAL FACILITY
1600 CALIFORNIA DR
VACAVILLE, CA 95687
707-448-6841 EXT 2002

9077
IRENE M. LOVEWELL
EVALUATIONS OFFICER
CALIFORNIA STATE POLYTECH UNIVERSITY
3801 WEST TEMPLE AVE
POMONA, CA 91767
714-869-3291

9076
EDWARD J. AUBERT
DIR OF ENROLLMENT SERVICES
CALIFORNIA STATE UNIVERSITY
801 WEST MONTE VISTA ST
TURLOCK, CA 95382
209-667-3256

*9078
CHERYL MADDERN
COORDINATOR
CALIFORNIA STATE UNIVERSITY DOMINGUEZ HILLS
1000 EAST VICTORIA ST
CARSON, CA 90747
310-243-3909

*8164
JAMES NEAL
DIR. TESTING & EVAL. SVCS
CALIFORNIA STATE UNIVERSITY LONG BEACH
1250 BELLFLOWER BLVD
LONG BEACH, CA 90840-0110
310-985-8385

8444
DON MOURTON
DEAN OF STUDENT SERVICES
CERRO COSO COMM COLLEGE
3000 COLLEGE HEIGHTS BLVD
RIDGECREST, CA 93555
619-375-5001

8662
CATHY BURKHART
DEAN, STUDENT SERVICES
CHAFFEY COLLEGE
5885 HAVEN AVE
RANCHO CUCAMONGA, CA 91701
714-987-1737

9092
SASKIA KNIGHT
REGISTRAR
CHAPMAN UNIVERSITY
333 NORTH GLASSELL ST
ORANGE, CA 92866
714-997-6815

This directory lists colleges & universities that award credit for passing scores on DSSTs.

*** INSTITUTIONS THAT ADMINISTER DSSTs**

**** Some schools may restrict testing to students admitted to or enrolled in the institution or offer a limited testing schedule.**

Students should contact the test center to determine current policies and schedules for DANTES testing.

8702
DR. BONNIE L. ROSS
ADVISOR, ACADEMIC CENTER
CHAPMAN UNIVERSITY
PO BOX 5818
VANDENBERG AFB, CA 93437

***9079**
STEVE PERDUE & CAROLE BAILEY
ASST. DIRECTORS
CHRISTIAN HERITAGE COLLEGE
2100 GREENFIELD DR
EL CAJON, CA 92019
619-442-5462

9103
CAROL BARNES
ASST. DEAN OF GUIDANCE
COASTLINE COMMUNITY COLLEGE
11460 WARNER AVE
FOUNTAIN VALLEY, CA 92708
714-546-7600

9865
TIM HARRINGTON
DIR, OUTREACH PROGRAM
COGSWELL POLYTECHNICAL COLLEGE
1175 BORDEAUX DR
SUNNYVALE, CA 95014
408-541-0150

***9474**
KARL ANDERSON
DIRECTOR
CONTINUING EDUCATION & ECONOMIC DEVELOPMENT
COLLEGE OF THE REDWOODS
7351 TOMPKINS HILL ROAD
EUREKA, CA 95501
707-445-6915

***8694**
DENISE VALDIVIA
SUPERVISOR OF TESTING
ETS-BAY AREA OFFICE
PO BOX 23060
OAKLAND, CA 94607
510-873-8086

9269
RONALD E. SHEPPARD
CENTER DIRECTOR
EMBRY-RIDDLE AERONAUTICAL UNIVERSITY
5146 ARNOLD AVE, SUITE 2
McCLELLAN AFB, CA 95652
916-920-9620

***9160**
PETER JOHNSON
ADMISSIONS ADMINISTRATOR
GOLDEN GATE UNIVERSITY
536 MISSION ST
SAN FRANCISCO, CA 94105
415-442-7203

9168
RON ESTROFF
REGISTRAR
HOLY NAMES COLLEGE
3500 MOUNTAIN BLVD
OAKLAND, CA 94619
415-436-1000

***8952**
LEANNE ENGKRAF-RIES
EDUCATIONAL TESTING PROGRAM
JOHN F. KENNEDY UNIVERSITY
12 ALTARINDA RD
ORINDA, CA 94563
925-258-2236

***8750**
VALERIE H. SMITH
DIRECTOR, DEVELOPMENTAL ED
LA SIERRA UNIVERSITY
4700 PIERCE ST
RIVERSIDE, CA 92505
909-785-2452

9204
WILLIAM F. THAYER
CHAIR, COUNSELING
LOS ANGELES CITY COLLEGE
855 NORTH VERMONT AVE
LOS ANGELES, CA 90029
213-669-4478

9219
TIM MAGUFFEE
PROGRAMS COORDINATOR, CAFB
MERCED COLLEGE
3600 M ST
MERCED, CA 95348
209-722-7893

9236
DAVID BURKE
DIRECTOR, FORT ORD CAMPUS
MONTEREY PENINSULA COLLEGE
980 FREMONT
MONTEREY, CA 93940
408-646-4242

***9242**
CINDY NICHOLAS
LEARNING RESOURCE CTR MGR
NATIONAL UNIVERSITY
4121 CAMINO DEL RIO SOUTH #17
SAN DIEGO, CA 92108-7296
619-563-7296

8958
GLORIA KERKHOFF
EVALUATIONS TECHNICIAN
PALOMAR COLLEGE
1140 W MISSION RD
SAN MARCOS, CA 92069-1487
619-744-1150 EXT 2165

***8810**
PAT ROBERTS
TEST COORDINATOR
STUDENT COUNSELING CENTER
PEPPERDINE UNIVERSITY
24255 PACIFIC COAST HWY
MALIBU, CA 90263
310-456-4210

***8325**
CATHY ALEXANDER
ASSMT TESTING COORDINATOR
SACRAMENTO CITY COLLEGE
3835 FREEPORT BLVD
SACRAMENTO, CA 95822
916-558-2540

***8426**
DIANE HANSEN
ASSOCIATE DIRECTOR OF ADMISSIONS
SAMUEL MERRITT COLLEGE
370 HAWTHRONE AVE
OAKLAND, CA 94609
510-869-6616

9297
SPENCER WATKINS
DEAN OF STUDENT SERVICES
SAN BERNARDINO VALLEY COLLEGE
701 SOUTH MT. VERNON AVE
SAN BERNARDINO, CA 92410
714-888-6511

9298
ATTN: TRANSCRIPT OFFICE
SAN DIEGO COMMUNITY COLLEGES
3375 CAMINO DEL RIO SOUTH
SAN DIEGO, CA 92108
619-584-6500

8980
DEANNA T. WONG
ENROLLMENT SERVICES
SAN FRANCISCO STATE UNIVERSITY
1600 HOLLOWAY AVE, ADM 154
SAN FRANCISCO, CA 94132
415-338-2141

***9273**
MERILYN L. HARGIS
ASSISTANT DIRECTOR
EXCEL DEGREE COMPLETION PROGRAM
SAN JOSE CHRISTIAN COLLEGE
790 SOUTH 12TH STREET
SAN JOSE, CA 95112
408-971-3923

8378
RAY LOU
ASSOC. V.P., CURRICULUM & ASMT
SAN JOSE STATE UNIVERSITY
ONE WASHINGTON SQUARE
SAN JOSE, CA 95192
408-924-2447

9314
JAMES BRACY
ASST DEAN OF STUDENT SVCS
SOLANO COMMUNITY COLLEGE
4000 SUISUN VALLEY RD
SUISUN CITY, CA 94585
707-864-7102

9436
NORMAN HETLAND
COOPERATIVE EXTENSION
UNIVERSITY OF CALIFORNIA
324 GRANNINI HALL
BERKELEY, CA 94720
415-642-6000

8153
ERIC GRAVENBERG
DIRECTOR OF ADMISSIONS
UNIVERSITY OF CALIFORNIA
900 UNIVERSITY AVE
RIVERSIDE, CA 92521
714-787-1012

This directory lists colleges & universities that award credit for passing scores on DSSTs.

*** INSTITUTIONS THAT ADMINISTER DSSTs**

**** Some schools may restrict testing to students admitted to or enrolled in the institution or offer a limited testing schedule.**

Students should contact the test center to determine current policies and schedules for DANTES testing.

*9353
MARILYN DAVIES
REGISTRAR
UNIVERSITY OF LA VERNE
1950 THIRD ST
LA VERNE, CA 91750
909-593-3511

*8457
KITT K. VINCENT
DIRECTOR
UNIVERSITY OF LA VERNE
PO BOX 5578
V.A.F.B, CA 93437
805-734-1306

*8741
MARCY S. COFFEE
TESTING COORDINATOR
UNIVERSITY OF PHOENIX
4900 CALIFORNIA AVE, TOWER A,
BAKERSFIELD, CA 93309
805-633-0300

*9458
LOIS RICHARDSON
RECEPTIONIST
UNIVERSITY OF PHOENIX
370 CHADBOURNE ROAD
FAIRFIELD, CA 94533
707-426-9500

*9461
CARLINA KEETER
TEST/ADMISSIONS COORDINATOR
UNIVERSITY OF PHOENIX
75 ROWLAND WAY, STE 100
NOVATO, CA 94945-5001
415-898-4449, X104

*9460
REBECCA RODRIGUEZ
UNIVERSITY OF PHOENIX
4643 QUAIL LAKES DRIVE
SUITE #202
STOCKTON, CA 95207
209-472-9980

*8048
NICOLE BAPTISTA
TESTING ADMNISTRATOR
UNIVERSITY OF PHOENIX
SUITE 100
1760 CREEKSIDE OAKS DR
SACRAMENTO, CA 95833
916-923-3914

*8751
JEANETTE SPAIN
TESTING COORDINATOR
UNIVERSITY OF PHOENIX
3870 MURPHY CANNON RD
SAN DIEGO, CA 92123
619-576-7469, X191

*8518
TESTING COORDINATOR
UNIVERSITY OF PHOENIX
185 BERRY ST, LOBBY 3
SAN FRANCISCO, CA 94107
415-495-3370

*9958
STEVE SCHUETT
TESTING DIRECTOR
UNIVERSITY OF PHOENIX
3590 NORTH 1ST STREET
SAN JOSE, CA 95134
408-435-8500

*8353
LYNN PERRY
TESTING COORDINATOR
UNIVERSITY OF PHOENIX
7901 STONERIDGE DRIVE
PLEASANTON, CA 94588
925-847-7640

*8254
LYNN PERRY
TESTING COORDINATOR
UNIVERSITY OF PHOENIX - WALNUT CREEK
370 NORTH WIGET LANE
WALNUT CREEK, CA 94598
925-847-7640

*8368
DONNA ARAVANIS
TESTING COORDINATOR
UNIVERSITY OF PHOENIX - FRESNO LEARNING CENTER
2490 WEST, SHAW AVE, SUITE #105
FRESNO, CA 93711
559-446-2660, X116

8345
ROGER A. MILLER
DEAN OF ADMISSIONS & REGISTRAR
WEST COAST UNIVERSITY
440 SHATTO PLACE
LOS ANGELES, CA 90020-1765
213-275-8828

*8140
KIM VAREY
ACADEMIC ADVISOR
WEBSTER UNIVERSITY
SAN DIEGO METROPOLITAN CAMPUS
6480 WEATHERS PLACE, SUITE 104
SAN DIEGO, CA 92121-3911
619-458-9310

*9978
MONICA PECOT
ASSOC DIRECTOR, CAREER SVCS
WOODBURY UNIVERSITY
7500 GLEN OAKS BLVD
BURBANK, CA 91504
818-767-0888

COLORADO

*8826
DARLENE NOLD
PROGRAM DIRECTOR OF ASSESSMENT
AIMS COMMUNITY COLLEGE
5401 WEST 20th ST
GREELEY, CO 80632
970-330-8008 EXT 299

*8797
CONNIE HUNT
TESTING COORDINATOR
ARAPAHOE COMMUNITY COLLEGE
2500 W. COLLEGE DR-P.O. BOX 9002
LITTLETON, CO 80160-9002
303-797-5659

9451
KELLY A. HUNT
DIRECTOR OF UNDERGRADUATE PROGRAMS
COLORADO CHRISTIAN UNIVERSITY - COLORADO SPRINGS CAMPUS
SCHOOL OF PROFESSIONAL STUDIES
685 CITADEL DRIVE EAST STE 225
COLORADO SPRINGS, CO 80909
719-573-6166

8687
LISA FRY
DIRECTOR OF ACADEMIC ADVISING
COLORADO CHRISTIAN UNIVERSITY GRAND JUNCTION CENTER
743 HORIZON COURT, SUITE 220
GRAND JUNCTION, CO 81506
970-242-1811

*9960
MICHAEL CHILDS
ACADEMIC ADVISORASSOCIATE REGISTRAR
COLORADO CHRISTIAN UNIVERSITY
180 SOUTH GARRISON ST
LAKEWOOD, CO 80226-7499
303-963-3268

*8615
LARRY DUTMER
TESTING COORDINATOR
COLORADO MOUNTAIN COLLEGE
1310 WESTHAVEN DRIVE
VAIL, CO 81657
970-476-4040

*8420
DONNA DAUS
STUDENT SERVICES COUNSELOR
COLORADO MOUNTAIN COLLEGE
1402 BLAKE AVE
GLENWOOD SPRINGS, CO 81601
970-945-7486

*9645
DAN SCHAFFRICK
STUDENT SERVICES COORD.
COLORADO MOUNTAIN COLLEGE
1330 BOB ADAMS DRIVE
STEAMBOAT SPRINGS, CO 80487
970-870-4455

*8922
SHEILA HARPER, ACTG DIR
ADULT LEARNING CENTER
COLORADO NORTHWESTERN COMMUNITY COLLEGE
50 COLLEGE DR
CRAIG, CO 81625
970-824-1126

8150
WILLIAM YOUNG
DIRECTOR OF ADMISSIONS
COLORADO SCHOOL OF MINES
GOLDEN, CO 80401
303-273-3000

*9111
SANDI MILLER
DIRECTOR OF ADMISSIONS
COLORADO TECHNICAL COLLEGE
4435 NORTH CHESTNUT ST
COLORADO SPRINGS, CO 80907
719-598-0200

*8825
THERESA CAMPBELL ARON
INTERIM DIR., LRNG RESOURCE CTR
LEARNING RESOURCES CENTER
COMMUNITY COLLEGE OF AURORA
16000 EAST CENTRETECH PKWY
AURORA, CO 80011-9036
303-360-4736

8972
ELAINE RYAN, COORDINATOR
CREDIT FOR PRIOR LEARNING
COMMUNITY COLLEGE OF DENVER
CAMPUS BOX 203, PO BOX 173363
DENVER, CO 80217
303-556-2481

*9482
DAN BURKE
ASSESSMENT COORDINATOR
DELTA MONTROSE AREA
VOCATIONAL TECHNICAL CENTER
1765 U.S. HWY 50
DELTA, CO 81416
970-874-7671

*9152
ALLYN TALG
DIRECTOR, CAREER SERVICES
FORT LEWIS COLLEGE
1000 RIM DR
DURANGO, CO 81301
970-247-7562

8003
LOUCAROL SHERIDAN, COORD
CREDIT FOR PRIOR LEARNING
FRONT RANGE COMMUNITY
COLLEGE
3645 WEST 112TH AVE
WESTMINSTER, CO 80030
303-466-8811

8341
GLORIA BROWN
ADMISSIONS COORDINATOR
INTERNATIONAL UNIVERSITY
PO BOX 6512
ENGLEWOOD, CO 80155-6512
303-784-8048

*8062
TERRI WINDOLPH
TESTING COORDINATOR
MESA STATE COLLEGE
1175 TEXAS AVE
GRAND JUNCTION, CO 81501
970-248-1215

9445
JAMES MORFELD
ASSOC DIR OF ADMISSIONS
METROPOLITAN STATE COLLEGE
PO 173362 CAMPUS BOX. 16
DENVER, CO 80217-3362
303-556-6970

*8832
DIANE PFEIFER
TESTING CENTER
MORGAN COMMUNITY COLLEGE
17800 ROAD 20
FORT MORGAN, CO 80701
970-542-3188

*8301
PETER HURST
ADMISSIONS OFFICE
NAROPA INSTITUTE
2130 ARAPAHOE AVE
BOULDER, CO 80302
303-444-0202

*8769
JUDY GIACOMINI
DIRECTOR, COUNSELING
NORTHEASTERN JR. COLLEGE
100 COLLEGE DR
STERLING, CO 80751
970-521-6657

*9268
DAVID KUEBBELER
COUNSELING SERVICES
OTERO JUNIOR COLLEGE
LA JUNTA, CO 81050
303-384-8721

*9279
BETTY ARMSTRONG
TEST CENTER COORDINATOR
PIKES PEAK COMMUNITY COLLEGE
5675 SOUTH ACADEMY BLVD
COLORADO SPRINGS, CO 80906
719-540-7130

*8599
VERA ESTRADA
DIRECTOR, OUTREACH SERVICES
PUEBLO COMMUNITY COLLEGE
900 WEST ORMOND AVE
PUEBLO, CO 81004
719-549-3200

*8277
DARYL D. YARROW
REGIS UNIVERSITY AT CMC
1402 BLAKE AVE
GLENWOOD SPRINGS, CO 81601-3961
970-945-0624

*8278
DIANE FREEMAN
REGIS UNIVERSITY NORTHEASTERN
JR. COLLEGE
100 COLLEGE DR
STERLING, CO 80751-2399
970-522-6600 EXT 7640

*9938
MELISSA FOSTER
SUPERVISOR OF TESTING SVCS
REGIS UNIVERSITY
3333 REGIS BLVD
DENVER, CO 80221
800-967-3237 EXT. 3575 OR
303-458-3575

9364
BARBARA NEIBERG
MBA PROGRAM DIRECTOR
UNIVERSITY OF COLORADO AT
COLORADO SPRINGS
PO BOX 7150
1420 AUSTIN BLUFFS PARKWAY
COLORADO SPRINGS, CO 80933-7150
719-262-3401

9448
PATRICIA JIMENEZ
ASST.DIRECTOR, ADMISSIONS
UNIVERSITY OF COLORADO
1200 LARIMER ST - CAMPUS BOX 167
DENVER, CO 80204
303-556-2659

CONNECTICUT

*8247
CATHERINE CRAWFORD
EDUCATIONAL ADVISOR
ALBERTUS MAGNUS COLLEGE
ONE LONG WHARF, SUITE 216
NEW HAVEN, CT 06511
800-394-9982

8645
HAKIM SALAHU-DIN
DIRECTOR OF ADMISSIONS
CENTRAL CONNECTICUT
STATE UNIVERSITY
1615 STANLEY ST
NEW BRITAIN, CT 06050
203-827-7000

*9928
PATRICIA DERECH
REGISTRAR
CHARTER OAK STATE COLLEGE
55 PAUL MANAFORT DRIVE
NEW BRITAIN, CT 06053
860-832-3820

9453
ARTHUR C. FORST, JR.
DIRECTOR OF ADMISSIONS
EASTERN CONNECTICUT
STATE UNIVERSITY
HURLEY HALL
WILLIMANTIC, CT 06226
203-456-2231

8756
DAVID FLYNN
DEAN OF ADMISSIONS
FAIRFIELD UNIVERSITY
FAIRFIELD, CT 06430
203-254-4000

8629
KATHERINE SAMSON
CONTINUING EDUCATION
HARTFORD COLLEGE FOR WOMEN
1265 ASYLUM AVE
HARTFORD, CT 06105
203-236-1215

This directory lists colleges & universities that award credit for passing scores on DSSTs.

*** INSTITUTIONS THAT ADMINISTER DSSTs**

**** Some schools may restrict testing to students admitted to or enrolled in the institution or offer a limited testing schedule.**

Students should contact the test center to determine current policies and schedules for DANTES testing.

*8777
ARTHUR BIRCH
ELECTRICAL ENGR.DEPT.
HARTFORD STATE TECHNICAL COLLEGE
401 FLATBUSH AVE
HARTFORD, CT 06106
203-527-4111

*8405
DIRECTOR OF TESTING
HOLY APOSTLES COLLEGE
33 PROSPECT HILL RD
CROMWELL, CT 06416
860-632-3033

*9454
VERONICA MALLORY
COORD. TUTORING & TESTING SVCS.
HOUSATONIC COMMUNITY-TECHNICAL COLLEGE
900 LAFAYETTE BLVD
BRIDGEPORT, CT 06604-4704
203-332-5217

9455
JOSEPH MESQUITA
DIRECTOR OF ADMISSIONS
MANCHESTER COMMUNITY TECHNICAL COLLEGE
60 BIDWELL ST
MANCHESTER, CT 06040
860-230-1717

*9475
BLANCA M. HERRERA
COORDINATOR OF TESTING
NAUGATUCK VALLEY COMMUNITY - TECHNICAL COLLEGE
750 CHASE PARKWAY
WATERBURY, CT 06708
203-575-8048

9261
CAROLYN THOMAS
TRANSFER COUNSELOR
NORWALK COMMUNITY COLLEGE
188 RICHARDS AVE
NORWALK, CT 06854-1655
203-857-7092

*8380
ELLEN E. COREY
TEST ADMINISTRATOR
SACRED HEART UNIVERSITY
5151 PARK AVE
FAIRFIELD, CT 06432-1000
203-396-8065

9459
MARILYN NIELSEN
CONTINUING EDUCATION
SAINT JOSEPH COLLEGE
1678 ASYLUM AVE
WEST HARTFORD, CT 06117
203-232-4571

*9281
RUTH SPRAGUE
DIRECTOR OF ADVISING
TEIKYO-POST UNIVERSITY
800 COUNTRY CLUB RD - BOX 2540
WATERBURY, CT 06723-2540
203-596-4623

*9233
CAROL E. KASZUBSKI
DIRECTOR OF TESTING
THREE RIVERS COMMUNITY TECHNICAL COLLEGE
MAHAN DR
NORWICH, CT 06360
860-823-2827

*9034
CHRISTINE GRANT
ADULT ACADEMIC SERVICES
UNIVERSITY OF HARTFORD
200 BLOOMFIELD COMP. CTR/231
WEST HARTFORD, CT 06112
860-768-4457

9949
JOSEPH P. MACIONUS
REGISTRAR
UNIVERSITY OF NEW HAVEN
300 ORANGE AVE
WEST HAVEN, CT 06516
203-932-7305

8864
MAUREEN O'HARA
ASSOC DIRECTOR OF ADMISSIONS
WESTERN CONNECTICUT STATE UNIVERSITY
181 WHITE ST
DANBURY, CT 06810
203-797-4347

8068
DONNA B. BIGHINATTI
REGISTRAR
WILCOX COLLEGE OF NURSING
28 CRESCENT ST
MIDDLETOWN, CT 06457

DELAWARE

*9464
JAMES F. HERTZOG
ASSOCIATE DIRECTOR OF TESTING
DELAWARE STATE UNIVERSITY
1200 NORTH DUPONT HIGHWAY
DOVER, DE 19901-2277
302-739-3581

*8594
D'MONT STITH
PRIOR LEARNING COORDINATOR
SPRINGFIELD COLLEGE
501 SHIPLEY ST
WILMINGTON, DE 19801
800-506-9270

DISTRICT OF COLUMBIA

9158
DAVID NEIMAN
ASST DIRECTOR OF ADMISSIONS
GEORGE WASHINGTON UNIVERSITY
2121 EYE ST, NW
WASHINGTON, DC 20052
800-447-3765

8573
WILLIAM SHERRILL
DEAN, ADMISSIONS & RECORDS
HOWARD UNIVERSITY
2400 SIXTH ST NW
WASHINGTON, DC 20059
202-806-6100

9442
THOMAS W. VAN WAGNER
ADMISSIONS OFFICER
JOINT MILITARY INTELLIGENCE COLLEGE
200 MACDILL BLVD, BOLLING AFB
WASHINGTON, DC 20340-5100
202-231-3299

*8855
ERVAN PEARSON
POTOMAC COLLEGE
4000 CHESAPEAKE STREET NW
WASHINGTON, DC 20016
202-686-0876

8351
JACK R. JONES, ED. D.
DEAN, INSTITUTIONAL PLANNING
STRAYER COLLEGE
1025 15th ST NW
WASHINGTON, DC 20005
202-408-2400

*8796
ANDREA LUKE
ENROLLMENT SERVICES
TRINITY COLLEGE
125 MICHIGAN AVE, NE
WASHINGTON, DC 20017
202-939-5040

FLORIDA

9523
MADELEINE G. WHITTAKER
DIR. TRANSCRIPT EVALUATION
BARRY UNIVERSITY
11300 N.E. 2nd AVE
MIAMI SHORES, FL 33161
305-899-3143

*8729
DONALD J. PITCHER
DIRECTOR OF TESTING
BETHUNE COOKMAN COLLEGE
640 SECOND AVE
DAYTONA BEACH, FL 32115
904-255-1401 EXT 409

9070
GAIL CHARK
COORDINATOR, CAEL AND CO-OP
BREVARD COMMUNITY COLLEGE
1519 CLEARLAKE RD SC215
COCOA, FL 32922
407-632-1111

*8816
THOMAS PERKINS
ASSSESSMENT SPECIALIST
BREVARD COMMUNITY COLLEGE
3865 NORTH WICKHAM RD (CM-209)
MELBOURNE, FL 32935
407-632-1111 EXT 32750

*8717
TERRY DOUGHERTY
TEST ADMINISTRATOR
TESTING CENTER
CENTRAL FLORIDA COMM COLL
OCALA, FL 33478
325-237-2111 EXT 332

*9551
JAMES FRASIER
DEAN, SPECIAL PROGRAMS
COMPUTER BASED TESTING CENTER
ECKERD COLLEGE
SPECIAL PROGRAMS DIVISION
4200 54TH AVENUE SOUTH
ST PETERSBURG, FL 33711
727-864-8241

8604
SUSAN POTTS
TEST ADMINISTRATOR
BARBARA SLAYBAUGH
TESTING SPECIALIST
EDISON COMMUNITY COLLEGE
8099 COLLEGE PKWY SW
FT. MYERS, FL 33906
941-489-9223

9143
PAMELA ARNETTE THOMAS
DIR OF ADMISS AND REGISTRAR
EMBRY-RIDDLE AERO UNIVERSITY
600 SOUTH CLYDE MORRIS BLVD
DAYTONA BEACH, FL 32114
904-226-7610

*8621
JEAN BAKER
ACADEMIC ASSESSMENT CENTER
FLORIDA COMMUNITY COLLEGE
KENT CAMPUS
3939 ROOSEVELT, BLDG A, RM 104
JACKSONVILLE, FL 32205
904-381-3491

9535
KATE ELLISON
ADMISSIONS ADMINISTRATOR
FLORIDA INSTITUTE OF TRADITIONAL CHINESE MEDICINE
5335 66TH STREET NORTH
ST. PETERSBURG, FL 33709
727-546-6565
800-565-1246

*8524
CAROLYN STEWART-CHILDERS
TESTING COORDINATOR
FLORIDA KEYS COMMUNITY COLLEGE
5901 COLLEGE RD
KEY WEST, FL 33040
305-296-9081, X237

*9466
KIMBERLY ELMHORST
DIRECTOR OF TESTING SERVICES
FLORIDA SOUTHERN COLLEGE
ORDWAY BUILDING - RM 142
McDONALD ST
LAKELAND, FL 33803
941-680-4299

9162
ANNE McCULLEN
DEAN OF STUDENT SERVICES
GULF COAST COMM COLLEGE
WEST HWY 98
PANAMA CITY, FL 32401
904-872-3871

*8753
ARTHUR OGDEN
CHAIR, LIBERAL ARTS PROGRAM
INTERNATIONAL COLLEGE OF NAPLES
2654 EAST TAMIAMI TRAIL
NAPLES, FL 33962
813-774-4700

9945
MARCIA L. GROSS
COORDINATOR, TRANSFER & OFF-CAMPUS SERVICES
JACKSONVILLE UNIVERSITY
2800 UNIVERSITY BLVD, NORTH
JACKSONVILLE, FL 32211
904-744-3950

*8731
GAYLE GARCIA
TESTING, RM 1303-7
MIAMI-DADE COMMUNITY COLLEGE-MEDICAL CTR CAMPUS
950 NW 20th ST
MIAMI, FL 33127
305-237-4359

9223
SAMUEL D. LAROUE,JR.
DIRECTOR OF ADMISSIONS
MIAMI-DADE COMMUNITY COLLEGE
300 N.E. SECOND AVE #5520
MIAMI, FL 33132-2297
305-237-7478

**9047
SUZAN KING
ACADEMIC SERVICES COUNSELOR
NATIONAL-LOUIS UNIVERSITY
604 COURTLAND, SUITE 150
ORLANDO, FL 32804
407-629-1000

**8854
DEANA SAWDEY
ACADEMIC SERVICES COUNSELOR
NATIONAL LOUIS UNIVERSITY
4890 W KENNEDY BLVD, SUITE 100
TAMPA, FL 33609-1836
813-286-8087

*9876
JUAN RAMIREZ
TESTING COORDINATOR
NOVA SOUTHEASTERN UNIVERSITY
3301 COLLEGE AVE
FORT LAUDERDALE, FL 33314
954-262-8370/8374

8862
CHRISTINE C. BISHOP
DIRECTOR FOR ENROLLMENT SVCS
REGISTRAR
OKALOOSA-WALTON COMMUNITY COLLEGE
100 COLLEGE BLVD
NICEVILLE, FL 32578
904-678-5111

8306
JAN BARNES
DIRECTOR, STUDENT PLACEMENT
ORLANDO COLLEGE
5500 DIPLOMAT CIRCLE
ORLANDO, FL 32810
407-628-5870

9655
CYNDEE DERENGOWSKI
STUDENT SERVICES COORD/PROGRAM ADVISOR
ROLLINS COLLEGE BREVARD
485 S. JOHN RODES BLVD
W. MELBOURNE, FL 32904
407-726-0432

8290
MARLEEN W. WIERSUM
ACADEMIC ADVISOR
SAINT LEO COLLEGE
3001 SW COLLEGE RD
PO BOX 2858 PORTABLE 4
OCALA, FL 34478
352-237-6922

9293
KAREN HATFIELD
REGISTRAR
OFFICE OF THE REGISTRAR - MC 2279
SAINT LEO COLLEGE
PO BOX 6665
ST. LEO, FL 33574-6665
352-588-8235

*9470
DR. PATRICIA SMITTLE
INSTITUTIONAL TEST ADMINISTRATOR
SANTA FE COMMUNITY COLLEGE
3000 NW 83rd ST
GAINSVILLE, FL 32606
904-395-5446

*8624
DR. DAVID GLAISE
COORDINATOR
SEMINOLE COMMUNITY COLLEGE
ASSESSMENT & TESTING CENTER
100 WELDEN BLVD
SANFORD, FL 32773
407-323-1450

8643
ROYCE M. SHELTON
DIRECTOR OF ADMISSIONS
SOUTHEASTERN COLLEGE
1000 LONGFELLOW BLVD
LAKELAND, FL 33801
813-665-4404

9096
LEONARD CHOJNACKI
DIRECTOR, HURLBURT/EGLIN
SAINT LEO COLLEGE
PO BOX 1835 WE AVENUE
BUILDING 252 113 A
EGLIN AFB, FL 32542
904-581-3755

This directory lists colleges & universities that award credit for passing scores on DSSTs.

*** INSTITUTIONS THAT ADMINISTER DSSTs**

**** Some schools may restrict testing to students admitted to or enrolled in the institution or offer a limited testing schedule.**

Students should contact the test center to determine current policies and schedules for DANTES testing.

9377
KAREN AUTREY
ASSISTANT DIRECTOR
SAINT LEO COLLEGE
NORTH FLORIDA CENTER, SANTA FE CC
3000 NW 83 ST, BLDG R RM 227
GAINESVILLE, FL 32606
352-395-5925

9365
JAMES OLNEY
DIRECTOR, KEY WEST
SAINT LEO COLLEGE
PO BOX 9033
NAS BUILDING #718 A
NAS KEY WEST, FL 33040
305-293-2848

9378
STEPHEN HESS
ASSISTANT DIRECTOR
SAINT LEO COLLEGE
NORTH FLORIDA CENTER
RT 19, BOX 1030, U.S. 90
LAKE CITY, FL 32025-8307
904-752-6866

8387
JANET C. MOORE
DIRECTOR
SAINT. LEO COLLEGE
PO BOX 6063
MACDILL AFB, FL 33608
813-840-0259

9379
BOB KORMAN
ASSISTANT DIRECTOR, MAYPORT
SAINT LEO COLLEGE
MAYPORT NAVAL STATION
BLDG #460, BOX 280015
MAYPORT, FL 32228
904-249-0911

9369
HEATHER DECKER
ASSISTANT DIRECTOR, PALATKA
SAINT LEO COLLEGE
NORTH FLORIDA CENTER, ST. JOHNS RIVER CC
5001 SAINT JOHNS, BOX 28
PALATKA, FL 32177-3897
904-325-1477

9429
PHIL MCCLUNG
DIRECTOR, CAMPUS CENTER - CDL
SAINT LEO COLLEGE
PO BOX 6665 MC-2248
SAINT LEO, FL 33574
352-588-8236

8728
DR. NAOMI B. WILLIAMS
COLLEGE REGISTRAR
ST. PETERSBURG JUNIOR COLLEGE
PO BOX 13489
ST. PETERSBURG, FL 33733

8180
ARTHUR A. SEIDNER
ASSESSMENT COORDINATOR
TRINITY INTERNATIONAL UNIVERSITY
COLLEGE OF LIBERAL ARTS
EXCEL DEPT
330 BISCAYNE BLVD, SUITE 202
MIAMI, FL 33132
305-381-7415 EXT 143

*8397
VIRGINIA KEMPER
COORD, INST. EFFECTIVENESS
TROY STATE UNIVERSITY
FLORIDA REGION
81 BEAL PKWY SE BOX 2829
FT. WALTON BCH, FL 32548
850-301-2147

8428
MARY S. BOLLA
DIRECTOR OF ADMISSIONS
UNIVERSITY OF NORTH FLORIDA
4567 ST JOHNS BLUFF RD SOUTH
JACKSONVILLE, FL 32224
904-646-2666

*8272
THOMAS MORTON
ACADEMIC COUNSELOR
UNIVERSITY OF PHOENIX FLORIDA CAMPUS
2600 LAKE LUCIEN DR, SUITE 200
MAITLAND, FL 32751
407-667-0555

*9042
EDWARD JUMP
DIRECTOR, UNDERGRADUATE BUSINESS PROGRAMS
UNIVERSITY OF SARASOTA - TAMPA
410 WARE BLVD, SUITE 300
TAMPA, FL 33619
813-246-4419

*9478
CHRIS ZEMAN
TEST COORDINATOR
UNIVERSTIY OF SOUTH FLORIDA
SVC 2054
4202 EAST FOWLER AVE
TAMPA, FL 33620
813-974-5298

8678
DR. RONALD JACK
REGISTRAR
WARNER SOUTHERN COLLEGE
5301 US HWY 27 S
LAKE WALES, FL 33853
941-638-1426

9650
PATRICIA K. HOEBER
ASSISTANT REGISTRAR
UNIVERSITY OF TAMPA
401 W. KENNEDY BLVD
TAMPA, FL 33606
813-253-6251

*8994
JOANN McKENNA
DIRECTOR, ADULT EDUCATION
WEBBER COLLEGE
1201 NORTH SCENIC HWY
BABSON PARK, FL 33827
941-638-1431

GEORGIA

*9970
ROY GOLDEN
COUNSELING & TESTING
ALBANY JUNIOR COLLEGE
2400 GILLIONVILLE RD
ALBANY, GA 31707

*9046
LYNN BENSON
DIR, COUNSELING SVCS
ARMSTRONG ATLANTIC STATE UNIVERSITY
11935 ABERCORN ST
SAVANNAH, GA 31419
912-927-5271

8757
WALTER JONES
VP, ACADEMIC AFFAIRS
ATLANTA METROPOLITAN COLLEGE
1630 STEWARD AVE, SW
ATLANTA, GA 30310
404-756-4441

*9049
DR. ROBERT MAYS
DIRECTOR
AUGUSTA STATE UNIVERSITY
2500 WALTON WAY
AUGUSTA, GA 30904-2200
706-737-1471

*8588
HAROLD JOHNSON
CHIEF EXAMINER
AUGUSTA EDUCATIONAL TESTING CENTER
415 BOY SCOUT RD
AUGUSTA, GA 30909

*9021
DR. DOYLE MATHIS
VICE PRESIDENT
ACADEMIC AFFAIRS
BERRY COLLEGE
PO BOX 640
MOUNT BERRY, GA 30149
706-232-2216

*9069
ANN MARIE ONG
DIRECTOR OF CAREER SERVICES
BRENAU UNIVERSITY
ONE CENTENNIAL CIRCLE
GAINESVILLE, GA 30501
770-534-6130

*8336
CYNTHIA A. LEE
DIR, CAREER DEVELOPMENT CTR
COASTAL GEORGIA COMMUNITY COLLEGE
3700 ALTAMA AVE
BRUNSWICK, GA 31523
912-264-7220

9113
CARL WALLMAN
DIRECTOR OF ADMISSIONS
COLUMBUS COLLEGE
ALGONQUIN DR
COLUMBUS, GA 31993
404-568-2001

8925
CHARLES T. EDWARDS
REGISTRAR
DARTON COLLEGE
2400 GILLIONVILLE RD
ALBANY, GA 31707

9969
LEVY MOORE
DIRECTOR OF ADMISSIONS
EMANUEL COLLEGE
FRANKLIN SPRINGS, GA 30639
404-245-7226

*8692
PATRICIA BAKR
TEST SUPERVISOR
EDUCATIONAL TESTING SERVICE
1979 LAKESIDE PKWY, SUITE 400
TUCKER, GA 30084
404-934-0160

*8056
SHARON JONES
TESTING OFFICE
GEORGIA COLLEGE
CAMPUS PO BOX 74
MILLEDGEVILLE, GA 31061
912-453-5350

*9159
LINDA WHATLEY
REGISTRAR
GEORGIA MILITARY COLLEGE
201 EAST GREENE ST
MILLEDGEVILLE, GA 31061
912-453-3481

*9481
WILLIAM LUNCEFORD
DIRECTOR
MERCER UNIVERSITY
3001 MERCER UNIVERSITY DR
ATLANTA, GA 30341
404-986-3000

*8675
MELVIN MILLS
ELECTRONICS DEPARTMENT
MOULTRIE AREA TECH INSTITUTE
BOX 520
MOULTRIE, GA 31768

**8821
RON LIPSCOMB
DIRECTOR OF ACADEMIC SERVICES
NATIONAL LOUIS UNIVERSITY
1777 N.E. EXPRESSWAY, SUITE 250
ATLANTA, GA 30329
404-633-1223 EXT 234

*8304
VICTOR WOLSKI
DIRECTOR, COUNSELING
REINHARDT COLLEGE
RTE. 140
WALESKA, GA 30183
404-479-1454

9389
DENNIS MARTIN
DIRECTOR, ATLANTA
SAINT LEO COLLEGE
ARMY EDUCATION CENTER
1302 COBB STREET SW
FT. MCPHERSON, GA 30330-1073
404-755-5898

9386
RICHARD CABBAGE
DIRECTOR, FT. STEWART
SAINT LEO COLLEGE
ARMY ED CTR.
PO BOX 3430, BLDG 130
FT. STEWART, GA 31315
912-767-8545

9388
RON WEIHER
DIRECTOR, HUNTER AAF
SAINT LEO COLLEGE
PO BOX 42119
BLDG #1290
SAVANNAH, GA 31409
912-352-3417

*8171
ANN HOUSE
DIRECTOR OF SPECIAL PROGRAMS
SHORTER COLLEGE
315 SHORTER AVE
ROME, GA 30165-4298
706-233-7266

*8642
AMY BELLEW
STUDENT SERVICES COORDINATOR
SHORTER COLLEGE
1950 SPECTRUM CIRCLE, STE. A-305
MARIETTA, GA 30067
404-980-3417 EXT 6612

*9622
JIM FISHER
STUDENT SERVICES COORDINATOR
SCHOOL OF PROFESIONAL PROGRAMS
SHORTER COLLEGE
585 OLD NORCROSS RD, STE K
LAWRENCEVILLE, GA 30045
770-237-5288, X101

8648
FRANCES KAMINSKY
DEAN OF ACADEMIC AFFAIRS
SOUTH COLLEGE
709 MALL BLVD
SAVANNAH, GA 31406
912-651-8100

9315
JOHN SAYLOR
DIRECTOR OF COUNSELING
SOUTH GEORGIA COLLEGE
COLLEGE PARK DR
DOUGLAS, GA 31533
912-383-4244

*8942
SUE SANDIFAR
ACADEMIC SERVICES
THOMAS COLLEGE
1501 MILLPOND RD
THOMASVILLE, GA 31792
912-226-1621

8414
WAYNE VOTER
PROFESSOR
TROY STATE UNIVERSITY
PO BOX 2456
FT BENNING, GA 31905
404-689-4744

*8350
DR. DWAINE SKILES
COORDINATOR, GENERAL
STUDIES PROGRAM
VALDOSTA STATE COLLEGE
VALDOSTA, GA 31698
912-333-7834

HAWAII

9091
GAIL C. KOKI
REGISTRAR
CHAMINADE UNIVERSITY
3140 WAIALAE AVE
HONOLULU, HI 96816
808-735-4722

9167
HELEN G. CHAPIN
DEAN FOR SATELLITE PROGRAMS
HAWAII PACIFIC UNIVERSITY
1166 FORT ST, SUITE 200
HONOLULU, HI 96813
808-544-0200

*9487
CHARLES ANDERSON
SOCAD COORDINATOR
HONOLULU COMMUNITY COLLEGE
874 DILLINGHAM BLVD
HONOLULU, HI 96825
808-845-9401

8339
JERRY LAMB
ENROLLMENT SERVICES
UNIVERSITY OF HAWAII
KAPIOLANI
4303 DIAMOND HEAD RD
HONOLULU, HI 96816
808-734-9559

IDAHO

*9066
CLARE SPOOR
TESTING COORDINATOR
BOISE STATE UNIVERSITY
1910 UNIVERSITY DR
BOISE, ID 83725
208-385-1011

9489
LENA PAXTON
RECORDS/TRANSCRIPT
COLLEGE OF SOUTHERN IDAHO
PO BOX 1238
TWIN FALLS, ID 83303
208-733-9554

9173
JOHN BOBELL
OCCUPATIONAL & EDUCATIONAL
PROGRAMS
IDAHO STATE UNIVERSITY
CAMPUS BOX 8081
POCATELLO, ID 83209
208-236-0211

This directory lists colleges & universities that award credit for passing scores on DSSTs.

*** INSTITUTIONS THAT ADMINISTER DSSTs**

**** Some schools may restrict testing to students admitted to or enrolled in the institution or offer a limited testing schedule.**

Students should contact the test center to determine current policies and schedules for DANTES testing.

*8488
PAT MARTIN
DEPARTMENT OF COUNSELING
LEWIS CLARK STATE COLLEGE
8th AVE & 6th ST
LEWISTON, ID 83501
208-799-2211

ILLINOIS

9058
LEE KNOEBEL
REGISTRAR
BELLEVILLE AREA COLLEGE
2500 CARLYLE RD
BELLEVILLE, IL 62221
618-235-2700

*8691
PATRICIA VARNER
COMMUNITY EDUCATION
BLACK HAWK COLLEGE
1501 STATE HWY 78
KEWANEE, IL 61443
309-852-5671 EXT 222

*9064
THOMAS QUAYLE
DEAN
BLACK HAWK COLLEGE
6600 34th AVE
MOLINE, IL 61265
309-796-1311

*9094
SYLVIA BUSH
DIRECTOR, TESTING AND ASSESSMENT
CITY COLLEGES OF CHICAGO
226 W JACKSON BLVD RM 912
CHICAGO, IL 60606
312-553-2500

9957
ATTN: TRANSCRIPTS
CITY COLLEGES OF CHICAGO
UNIT 29651
APO, AE 09096
011-49-6134-26633

9107
EUGENE HALLONGREN
ASSOC VICE PRESIDENT
COLLEGE OF DUPAGE
425 22nd ST
GLEN ELLYN, IL 60137-6599
630-942-2800

*9109
CAROL CRANE
TESTING CENTER SUPERVISOR
COLLEGE OF LAKE COUNTY
19351 WEST WASHINGTON ST
GRAYSLAKE, IL 60030
847-223-6601 EXT 2443

*9546
GRACE BOOY
LRC COORDINATOR
LEARNING RESOURCE CENTER
COLLEGE OF LAKE COUNTY
LAKESHORE CAMPUS
33 NORTH GENESEE STREET
WAUKEGAN, IL 60085
847-543-2120

9133
DEBRA HOPGOOD
ACTING DIR ACADEMIC TESTING
EASTERN ILLINOIS UNIVERSITY
CHARLESTON, IL 61920
217-581-5986

*8695
OLIVIA ATTENBORO
TEST SUPERVISOR
EDUCATIONAL TESTING SERVICE
ONE ROTARY CENTER
1560 SHERMAN AVE, SUITE 300
EVANSTON, IL 60201
847-869-7700

9583
STEVE HOLLER
ACADEMIC DIRECTOR
GREENVILLE COLLEGE
315 EAST COLLEGE AVENUE
GREEENVILLE, IL 62246
618-664-2800

8947
ZEKE ZEWDE
DIRECTOR OF STUDENT AFFAIRS
HAROLD WASHINGTON COLLEGE
30 EAST LAKE STREET
CHICAGO, IL 60601
312-553-5918

*8639
CECILIA GLODEN
ACADEMIC ADVISOR
HIGHLAND COMMUNITY COLLEGE
2998 WEST PEARL CITY RD
FREEPORT, IL 61032
815-235-6121

*8707
JOAN A. BLEVINS
TESTING ASSISTANT
ADVSMENT/CNSLG/ASSMT OFFICE
ILLINOIS CENTRAL COLLEGE
EAST PEORIA, IL 61635
309-694-5234

8302
E. KENTON PEAK
DIRECTOR, EDUCATIONAL SVCS
ILLINOIS EASTERN COMMUNITY COLLEGE
233 EAST CHESTNUT
OLNEY, IL 62450
618-393-2982

*9500
KEITH BAILEY
ADULT EDUCATION
JOHN WOOD COMMUNITY COLLEGE
150 SOUTH 48th ST
QUINCY, IL 62301
217-224-6500

8959
J. D. ROSS
DEAN OF CAREER ED. AND
ASST VP FOR ACADEMIC AFFAIRS
JOLIET JUNIOR COLLEGE
1215 HOUBOLT AVE
JOLIET, IL 60431
815-773-6629

9443
JAYNN TOBIAS-JOHNSON
STUDENT SERVICES, DIVISION OF EDUCATION
JUDSON COLLEGE
1151 NORTH STATE STREET
ELGIN, IL 60123
847-695-2500 EXT 2234

9501
LEA HOUDEK
DIR, ADMISSIONS/REG/RECORDS
KISHWAUKEE COLLEGE
21193 MALTA RD
MALTA, IL 60150
815-825-2086 EXT 218

9502
LINDA McVEY
TESTING COORDINATOR
LEWIS & CLARK COMMUNITY COLLEGE
5800 GODFREY RD
GODFREY, IL 62035
618-466-3411

9217
SUE CORDON
DIRECTOR OF ADMISSIONS
McKENDREE COLLEGE
701 COLLEGE RD
LEBANON, IL 62254
618-537-4481

9505
ROD SEANEY
CTR. ALTERNATIVE LEARNING
MORAINE VALLEY COMMUNITY COLLEGE
10900 SOUTH 88th AVE
PALOS HILLS, IL 60465
312-974-4300

**8340
KATHI SKRUDLAND
ASSESSMENT COUNSELOR
NATIONAL-LOUIS UNIVERSITY
200 SOUTH NAPERVILLE RD
WHEATON, IL 60187-5422
630-668-3838 EXT 4315

9255
ULESTINE WATSON
NONTRADITIONAL DEGREE PROGRAM
NORTHEASTERN ILLINOIS UNIVERSITY
5500 NORTH SAINT LOUIS AVE
CHICAGO, IL 60641
312-794-2853

9262
TRUDY BERS
SENIOR DIRECTOR OF RESEARCH, CURRICULUM & PLANNING
OAKTON COMMUNITY COLLEGE
1616 EAST GOLF RD
DES PLAINES, IL 60016
708-635-1894

8403
BILL G. HUTCHERSON
ADMINISTRATOR
PARK COLLEGE RESIDENT CENTER
607 PIERCE ST, RM 119
SCOTT AFB, IL 62225-5421
618-744-1555

9275
DR. JOHN HEDEMAN
DIRECTOR OF COUNSELING
PARKLAND COLLEGE
2400 WEST BRADLEY ST
CHAMPAIGN, IL 61821-1899
217-351-2364

***8633**
PAMELA SMITH
DEAN, STUDENT ASSESSMENT
REND LAKE COLLEGE
INA, IL 62846
618-437-5321

***9992**
TEENA ZINDEL-McWILLIAMS
COORDINATOR OF TESTING
RICHLAND COMMUNITY COLLEGE
ONE COLLEGE PARK
DECATUR, IL 62521
217-875-7211 EXT 364

9014
ELIZABETH HOOVER
REGISTRAR
SCHOOL OF THE ART INSTITUTE OF CHICAGO
37 SOUTH WABASH AVE
CHICAGO, IL 60603-3103
312-899-5117

***8242**
DR. MICHAEL SCHNUR
TESTING COUNSELOR
SOUTH SUBURBAN COLLEGE
15800 SOUTH STATE ST
SOUTH HOLLAND, IL 60473
708-596-2000

9318
DR. PATRICIA COVINGTON
ASSOCIATE DIRECTOR OF ADMISSIONS AND RECORDS
ADMISSIONS AND RECORDS
MAIL CODE 4701
SOUTHERN ILLINOIS UNIVERSTIY
CARBONDALE, IL 62901
618-453-4381

9319
CHRISTA OXFORD
ADMISSIONS
SOUTHERN ILLINOIS UNIVERSITY
EDWARDSVILLE, IL 62026-1047
618-692-2080

***9343**
MARGARET WILL
TEST COORDINATOR
TRITON COMMUNITY COLLEGE
2000 FIFTH AVE
RIVER GROVE, IL 60171
708-456-0300

9513
JR. LEE HAMM
ADMISSIONS/RECORDS OFFICE III
WESTERN ILLINOIS UNIVERSITY
5 HORRABIN HALL
MACOMB, IL 61455
309-298-1929

INDIANA

9516
EVERETT KOHLS
DEAN OF STUDENT SERVICES
BUTLER UNIVERSITY
4600 SUNSET AVE
INDIANAPOLIS, IN 46208
317-283-8000

9996
BARBARA HOWALD
REGISTRAR
FRANKLIN COLLEGE OF INDIANA
501 EAST MONROE ST
FRANKLIN, IN 46131
317-736-8441

9995
JOHN NYCE
DEAN
GOSHEN COLLEGE
GOSHEN, IN 46526
219-535-7000

***8034**
KRIS CHAFIN
DIRECTOR OF LEARNING ASSISTANCE
HUNTINGTON COLLEGE
2303 COLLEGE AVE
HUNTINGTON, IN 46750-1299
219-359-4290 EXT 2090

9521
DON ST. CLAIRE
DIRECTOR, EXTENDED STUDIES
INDIANA INSTITUTE OF TECHNOLOGY
1600 EAST WASHINGTON BLVD
FORT WAYNE, IN 46803
219-422-5561

***8726**
LISA NALLY
COUNSELOR
INDIANA UNIVERSITY
2300 SOUTH WASHINGTON ST
KOKOMO, IN 46904
317-453-2000

8853
KAY O'BRIEN
CONTINUING STUDIES
INDIANA UNIVERSITY EAST
2325 CHESTER BLVD
RICHMOND, IN 47374
317-973-8312

***8562**
BOB LOVELY
CHAIR, CONTINUING STUDIES
INDIANA UNIVERSITY NORTHWEST
3400 BROADWAY
GARY, IN 46408
219-980-6829

***9556**
PATRICIA J. DAVIS
STUDENT RECORDS SPECIALIST
ADMINISTRATION BUILDING
INDIANA UNIVERSITY - SOUTH BEND
1700 MISHAWAKA AVENUE
P.O. BOX 7111
SOUTH BEND, IN 46634-7111
219-237-4260

***8867**
DOUG DENTON
STUDENT DEVELOPMENT CENTER
INDIANA UNIVERSITY SOUTHEAST
4201 GRANT LINE RD
KNOBVIEW 235
NEW ALBANY, IN 47150
812-941-2312

***8683**
BETSY KIGER
MANAGER OF EXTERNAL TESTING
IUPUI TESTING CENTER
INDIANA UNIVERSITY-PURDUE UNIVERSITY - INDIANAPOLIS
425 UNIVERSITY BLVD CA129
INDIANAPOLIS, IN 46202-5143
317-274-2620

8565
SHARON M. MARTIN
REGISTRAR
INDIANA VOCATIONAL TECHNICAL COLLEGE
3116 CANTERBURY CT
BLOOMINGTON, IN 47404

***8781**
DIANNA WILLIAMS
TESTING CENTER
INDIANA WESLEYAN UNIVERSITY
4201 SOUTH WASHINGTON ST
MARION, IN 46953
765-677-2257

***9316**
LAURIE TOBEN
ASSESSMENT COORDINATOR
IVY TECH STATE COLLEGE
1815 E. MORGAN
PO BOX 1373
KOKOMO, IN 46903-0561
765-459-0561 EXT 344

***8388**
MICHAEL FISCHER
ASSESSMENT COORDINATOR
IVY TECH STATE COLLEGE
7999 US HWY 41 SOUTH
TERRE HAUTE, IN 47802-4898
812-299-1121

9212
SR. CATHERINE RATERS
DIRECTOR OF TESTING
MARTIN UNIVERSITY
2171 AVONDALE PLACE
INDIANAPOLIS, IN 46218
317-543-3235

***8445**
JERRY W. PHILLIPS
DEAN OF ACADEMIC AFFAIRS
OAKLAND CITY UNIVERSITY
143 N. LUCRETIA ST
OAKLAND CITY, IN 47660-1099
812-749-1238

9328
ALICE QUINLAN
EXTERNAL DEGREE PROGRAM
SAINT MARY OF THE WOODS
ST. MARY WOODS, IN 47876
812-535-5151

*8930
DR. EDWARD MARTIN
ACADEMIC DEAN
TRINITY COLLEGE & SEMINARY
4233 MEDWEL DR
NEWBURGH, IN 47629-0717
812-856-0611

*9351
LYNN PENLAND
DIRECTOR
UNIVERSITY OF EVANSVILLE
1800 LINCOLN AVE
EVANSVILLE, IN 47711
812-479-2981

*9520
DIANE METHENY
REGISTRAR
UNIVERSITY OF INDIANAPOLIS
1400 EAST HANNA AVE
INDIANAPOLIS, IN 46227
317-788-3262

8564
MAUREEN MCCON
DIRECTOR, WEEKEND COLLEGE
UNIVERSITY OF SAINT FRANCIS
2701 SPRING ST
FORT WAYNE, IN 46808
219-434-3294

*9382
KAREN SUTTON McMAHAN
DANTES/CLEP EXAMINER
VINCENNES UNIVERSITY
1002 NORTH FIRST ST
VINCENNES, IN 47591
812-888-8888

9383
PAT BELL
FT. HARRISON CENTER
VINCENNES UNIVERSITY
BUILDING 43
FT. BENJAMIN HARRISON, IN 47591
317-542-4401

IOWA

*8714
SR. JEAN BERINGER
CONTINUING EDUCATION
BRIAR CLIFF COLLEGE
PO BOX 2100
SIOUX CITY, IA 51104
712-279-5321

*8316
LORI McKINLEY
TEST ADMINISTRATOR
GRADUATE & ADULT CONTINUING
EDUCATION STUDIES
CLARK COLLEGE
1550 CLARK DR
DUBUQUE, IA 52001-3198
319-588-6354

*8076
CAROLYN WASSENAAR
DIRECTOR, ACADEMIC SUPPORT
GRAND VIEW COLLEGE
1200 GRANDVIEW AVE
DES MOINES, IA 50316-1599
515-263-2971

8971
WILLIAM V. HANSEN
VETERANS COORDINATOR
INDIAN HILLS COMMUNITY COLLEGE
525 GRANDVIEW
OTTUMWA, IA 52501
515-683-5111

*9026
TOM SPALDING
REGISTRAR
IOWA LAKES COMMUNITY COLLEGE
300 SOUTH 18th
ESTHERVILLE, IA 51334
712-362-7923

8040
DAVID FILE
DIRECTOR, CONTINUING ED
IOWA WESLEYAN COLLEGE
601 NORTH MAIN ST
MT. PLEASANT, IA 52641
319-385-8021

9203
GERARD B. NOONAN
REGISTRAR
LORAS COLLEGE
1450 ALTA VISTA
DUBUQUE, IA 52001
319-588-7106

*8785
JEFF ARMSTRONG
DEAN OF THE COLLEGE
MUSCATINE COMMUNITY COLLEGE
152 COLORADO ST
MUSCATINE, IA 52761
319-263-8250

9030
CATHY WALKER-SEARCY
REGISTRAR
SCOTT COMMUNITY COLLEGE
500 BELMONT RD
BETTENDORF, IA 52722
319-441-4130

8586
MIRIAM BOWER BORR
REGISTRAR
SIMPSON COLLEGE
701 NORTH C ST
INDIANOLA, IA 50125
515-961-1642

*8835
MARILYN WERNER
ASSESSMENT CENTER
SOUTHWESTERN COMMUNITY COLLEGE
1501 WEST TOWNLINE
CRESTON, IA 50801
515-782-1330

*9531
STEPHANIE THOMPSON
CAREER SERVICES
TEIKYO-MARYCREST UNIVERSITY
1607 WEST 12th ST
DAVENPORT, IA 52804
319-326-9233

*8937
JULIE JOHNSON
EXAMINATION ADMINISTRATIONS
COORDINATOR
ACADEMIC ADVISING SVCS CENTER
UNIVERSITY OF NORTHERN IOWA
125 STUDENT SERVICES
CEDAR FALLS, IA 50614-0383
319-273-6024

9376
HOLLY STREETER
ASSOCIATE REGISTRAR
UPPER IOWA UNIVERSITY
PO BOX 1861
FAYETTE, IA 52142
319-425-5265

*8284
JULIE SCOTT
COORDINATOR FOR PRIOR LEARNING
WESTERN IOWA TECH. COMMUNITY COLLEGE
4647 STONE AVE, PO BOX 5199
SIOUX CITY, IA 51102-5199
712-274-8733 EXT 1207

*9642
BRENDA STEINKE
DIR., CAREER SERVICES
WILLIAM PENN COLLEGE
201 TRUEBLOOD AVE
OSKALOOSA, IA 52577
515-673-1041

KANSAS

9398
BARBARA LEAVITT
REGISTRAR
ALLEN COUNTY COMMUNITY COLLEGE
1801 NORTH COTTONWOOD
IOLA, KS 66749
316-365-5116

*8343
LOUIS PIVONKA
ASSESSMENT LEARNING CENTER
BARTON COUNTY COMMUNITY COLLEGE
RTE. 3
GREAT BEND, KS 67530
316-792-2701

*8176
STEPHEN HALLER
DIRECTOR
BROWN-MACKIE COLLEGE
126 SOUTH SANTA FE
SALINA, KS 67401
913-825-5422

9074
EVERETT KOHLS
DEAN OF STUDENT SERVICES
BUTLER COUNTY COMMUNITY COLLEGE
HAVERHILL RD/TOWANDA AVE
EL DORADO, KS 67042
316-321-5083

8730
DENNIS NEIL
ASSOCIATE DEAN
DODGE CITY COMMUNITY COLLEGE
1201 FIRST AVE
DODGE CITY, KS 67801
316-225-0123

*9151
CARLA HATTAN
ASST DIRECTOR, KELLY CENTER
FORT HAYS STATE UNIVERSITY
600 PARK ST
HAYS, KS 67601-4099
913-628-4000

*8791
DEBRA MARTIN
DIRECTOR
STUDENT DEVELOPMENT
FRIENDS UNIVERSITY
2100 UNIVERSITY
WICHITA, KS 67213
316-295-5832

*9999
BETH TEDROW
DEAN STUDENT SERVICES
GARDEN CITY COMMUNITY COLLEGE
801 CAMPUS DR
GARDEN CITY, KS 67846
316-276-7611

*9536
JOSEPH McCANN
DEAN OF INSTRUCTION
HIGHLAND COMMUNITY COLLEGE
BOX 68
HIGHLAND, KS 66035
913-442-3236

9172
JOEL KELLER
DIRECTOR OF RECORDS
HUTCHINSON COMMUNITY COLLEGE
1300 NORTH PLUM ST
HUTCHINSON, KS 67501
316-665-3500

9184
RICHARD LEE
DEAN OF STUDENT SERVICES
KANSAS CITY KANSAS COMMUNITY COLLEGE
7250 STATE AVE
KANSAS CITY, KS 66112
913-334-1100

*9186
RITA ROBERTS
COORDINATOR OF TESTING PROGRAMS
KANSAS STATE UNIVERSITY
HOLTON HALL, RM 101B
MANHATTAN, KS 66506
913-532-6492

8346
BILL R. POUNDS
DEAN OF STUDENT SERVICES
LABETTE COMMUNITY COLLEGE
200 SOUTH 14th ST
PARSONS, KS 67357
316-421-6700

*8875
SHARON REED
ASSISTANT DIRECTOR
LEAD DEGREE COMPLETION PROGRAM
MANHATTAN CHRISTIAN COLLEGE
1415 ANDERSON AVE
MANHATTAN, KS 66501-4081
785-539-4841

*8801
JERRI LYNN SAPP
ACADEMIC COORDINATOR
MHR PROGRAM
MIDAMERICA NAZARENE COLLEGE
2030 COLLEGE WAY
OLATHE, KS 66062
913-791-3277

9185
SHIRLEY RUEB
REGISTRAR
NEWMAN UNIVERSITY
3100 McCORMICK AVE
WICHITA, KS 67213
316-942-4291, X121

*8655
MARY A. JOHNSON
REGISTRAR
OTTAWA UNIVERSITY
10865 GRANDVIEW, SUITE 2000
OVERLAND PARK, KS 66210
913-451-0120 EXT12

8408
LINDA OCCHIUZZO
ADMINISTRATOR
PARK COLLEGE AT FORT LEAVENWORTH
ARMY ED CENTER, BLDG #74
425 McPHERSON AVE
FORT LEAVENWORTH, KS 66027
913-682-9229

*9538
DAMON BROADUS
TESTING
PITTSBURG STATE UNIVERSITY
OVERMAN STUDENT CENTER
PITTSBURG, KS 66762
316-235-4264

8337
CAROL HINDS
ACADEMIC DEAN
SAINT MARY COLLEGE
4100 SOUTH 4th ST, TRAFFICWAY
LEAVENWORTH, KS 66048
913-682-5151

*8876
TOM CORNELIUS
DIRECTOR, BUSINESS, INDUSTRY, & COMMUNITY SERVICE
SEWARD COUNTY COMMUNITY COLLEGE
1801 NORTH KANSAS
LIBERAL, KS 67905-1137
316-624-1954

*9534
BECKY A. KLASSEN
PROGRAM REPRESENTATIVE
DEGREE COMPLETION PROGRAMS
TABOR COLLEGE WICHITA
7348 W. 21ST ST., SUITE 117
WICHITA, KS 67205
316-773-9783

9543
NORMA HENLEY
TRANSCRIPT ANALYST
UNIVERSITY OF KANSAS
126 STRONG HALL,
LAWRENCE, KS 66045
913-864-5136

*9391
GAYLE VELTMAN
TESTING COORDINATOR
WICHITA STATE UNIVERSITY
1845 NORTH FAIRMOUNT
320 GRACE WILKIE HALL
WICHITA, KS 67260-0091
316-978-3440

KENTUCKY

*8577
WILLIE McCULLOUGH
DEAN OF STUDENT AFFAIRS
ASHLAND COMMUNITY COLLEGE
1400 COLLEGE DR
ASHLAND, KY 41101
606-329-2999

9544
ROBERT PFAADT
REGISTRAR
BELLARMINE COLLEGE
2001 NEWBURG RD
LOUISVILLE, KY 40205
502-452-8133

8830
SISTER JOANN MARK
ACADEMIC DEAN
BRESCIA COLLEGE
717 FREDERICA ST
OWENSBORO, KY 42301
502-686-4250

8005
SYLVIA MORRIS
DEAN, STUDENT DEVELOPMENT
CAMPBELLSVILLE COLLEGE
200 W COLLEGE ST
CAMPBELLSVILLE, KY 42718
502-465-8158

9134
JAMES L. GRIGSBY
DIRECTOR OF ADMISSIONS
EASTERN KENTUCKY UNIVERSITY
649 COATES
RICHMOND, KY 40475
606-622-2101

This directory lists colleges & universities that award credit for passing scores on DSSTs.

*** INSTITUTIONS THAT ADMINISTER DSSTs**
**** Some schools may restrict testing to students admitted to or enrolled in the institution or offer a limited testing schedule.**

Students should contact the test center to determine current policies and schedules for DANTES testing.

9169
FRANK MONTGOMERY
DIR, FT. CAMPBELL CAMPUS
HOPKINSVILLE COMMUNITY COLLEGE
PO BOX 2100
HOPKINSVILLE, KY 42241
615-431-3777

8979
CARRIE MORRIS
VETERANS CERTIFICATION SPECIALIST
KENTUCKY STATE UNIVERSITY
ACADEMIC SERVICES BLDG 323
FRANKFORT, KY 40601
502-227-6000

*9190
SCOTT E. KRAMER
REGISTRAR
KENTUCKY WESLEYAN COLLEGE
3000 FREDERICA ST
OWENSBORO, KY 42301
502-926-3111

*8813
BEVERLY BRYANT
TESTING ADMINISTRATOR
LINDSEY WILSON COLLEGE
210 LINDSEY WILSON ST
COLUMBIA, KY 42728
502-384-2126

8811
GENE OTTIE
ADMISSIONS/FINANCIAL AID
McKENDREE COLLEGE
1635 W LINCOLN TRAIL BLVD
RADCLIFF, KY 40160

*8334
JEFFREY STONE
DIRECTOR OF ADMISSIONS
MIDWAY COLLEGE
SCHOOL FOR CAREER DEVLMT.
512 EAST STEPHENS
MIDWAY, KY 40347
606-846-5800

*8274
DR. WILLIAM ALLBRITTEN
DIRECTOR, COUNSELING & TESTING CENTER
MURRAY STATE UNIVERSITY
PO BOX 9
MURRAY, KY 42072-0009
502-762-6851

8814
JOE YAZVAC
COUNSELOR
OWENSBORO COMMUNITY COLLEGE
4800 NEW HARTFORD RD
OWENSBORO, KY 42303
502-686-4416

*9542
DOROTHY THOMAS, SCN
DIRECTOR
CENTER FOR ACADEMIC DEVELOPMENT
SPALDING UNIVERSITY
851 SOUTH FOURTH STREET
LOUISVILLE, KY 40203
502-585-9911 EXT 276

8656
BARBARA DEAN
DIRECTOR OF ADMISSIONS
SULLIVAN COLLEGE
P.O BOX 245
FORT KNOX, KY 40121
502-942-8503

8494
CONNIE S. COARTNEY
REGISTRAR
SULLIVAN COLLEGE
3101 BARDSTOWN RD
LOUISVILLE, KY 40205
502-456-6504

*9550
MAUREEN BENSMAN
ACADEMIC ADVISOR
CONTINUING EDUCATION
THOMAS MORE COLLEGE
333 THOMAS MORE PARKWAY
CRESTVIEW HILLS, KY 41017
606-344-3333

*9906
SHARON ERCEY
COORDINATOR OF TESTING
WESTERN KENTUCKY UNIVERSITY
409 POTTER HALL
BOWLING GREEN, KY 42101

8543
MARTHA YOUNG
ADMISSIONS OFFICER
UNIVERSITY OF KENTUCKY
100 W.D. FUNKHOUSER BLDG
LEXINGTON, KY 40506
606-257-1806

*9548
LEA DAVIS
COORDINATOR OF TESTING SERVICES
UNIVERSITY OF LOUISVILLE
DAVIDSON HALL, RM 310
106 EAST BRANDEIS ST
LOUISVILLE, KY 40292
502-852-6606

*9906
SHARON URSEY
COORDINATOR OF TESTING
WESTERN KENTUCKY UNIVERSITY
409 POTTER HALL
BOWLING GREEN, KY 42101
502-745-3159

LOUISIANA

*9549
LEWIS BETTINGER
TESTING COORDINATOR
CENTENARY COLLEGE OF LOUISIANA
2911 CENTENARY BLVD
SHREVEPORT, LA 71134
318-869-5011

9207
VERONA SMITH
DIRECTOR, ORIENTATION & SPECIAL SERVICES
LOUISIANA COLLEGE
1140 COLLEGE DR
PINEVILLE, LA 71359
318-487-7135

9208
B. J. OWENS
OFFICE OF THE REGISTRAR
LOUISIANA TECH. UNIVERSITY
BOX 3155-TECH STATION
RUSTON, LA 71272
318-257-2176

*9270
GERALD F. DELUCA
ACADEMIC PRESIDENT
OUR LADY OF HOLY CROSS COLLEGE
4123 WOODLAND DR
NEW ORLEANS, LA 70131
504-394-7744

*9540
RAINY MENDOZA
DIRECTOR OF TESTING
OFFICE OF TESTING
SOUTHEASTERN LOUISIANA UNIVERSITY
BOX 750, SLU
HAMMOND, LA 70402
504-549-3897

*8483
W. C. ARMSTRONG
TESTING SERVICES
SOUTHERN UNIVERSITY
PO BOX 10777
BATON ROUGE, LA 70813
504-771-4500

*8763
CHARLES TREADWAY
TESTING AND ASSESSMENT
SOUTHERN UNIVERSITY
3050 MARTIN LUTHER KING DR
SHREVEPORT, LA 71107
318-674-3368

8157
CHERYL L. BERNIER
DIRECTOR, FORT POLK CENTER
UPPER IOWA UNIVERSITY
PO BOX 15015
FORT POLK, LA 71459
318-537-4465

MAINE

9054
LOUISE GRANT
ADMISSIONS ADMINISTRATOR
BEAL COLLEGE
629 MAIN ST
BANGOR, ME 04401
207-947-4591

9171
JOHN BENOIT
CONTINUING EDUCATION
HUSSON COLLEGE
ONE COLLEGE CIRCLE
BANGOR, ME 04401
207-947-1121

8448
RONDA HERBERT-GAGNE
ENROLLMENT MANAGEMENT
SPECIALIST (C-119)
KENNEBEC VALLEY TECHNICAL COLLEGE
92 WESTERN AVENUE
FAIRFIELD, ME 04937
800-528-5882

8663
WILLIAM C. BEIL
CONTINUING EDUCATION
NORTHERN MAINE TECHNICAL COLLEGE
33 EDGEMONT DR
PRESQUE ISLE, ME 04769
207-328-9037

*8668
ELAINE M. TRUMBLE
ASST DEAN OF OPERATIONS
DIV OF CONTINUING AND
PROFESSIONAL STUDIES
ST. JOSEPH'S COLLEGE
278 WHITES BRIDGE RD
STANDISH, ME 04084-5263
207-893-7804

9552
JANE MARTEL
DIRECTOR OF ADMISSIONS
THOMAS COLLEGE
180 WEST RIVER RD
WATERVILLE, ME 04901
207-877-0101

9355
CHARLES P. AMES
UNIVERSITY PROGRAMS
UNIVERSITY OF MAINE
181 MAIN
PRESQUE ISLE, ME 04769
207-328-6827

*8612
JOYCE BLANCHARD GARAND
COORD. TESTING SERVICES/LRNG LAB
UNIVERSITY OF MAINE
46 UNIVERSITY DR
AUGUSTA, ME 04330-9410
207-621-3495

8784
CYNTHIA KEMBLE
PLACEMENT COORDINATOR
UNIVERSITY OF MAINE
SOUTH ST
FARMINGTON, ME 04938
207-778-3501

*9354
GEORGE DIAZ
ACADEMIC COUNSELOR
UNIVERSITY OF MAINE
25 PLEASANT ST
FORT KENT, ME 04743
207-834-7532

*9553
KATHIE BLINICK
DIRECTOR
PRIOR LEARNING ASSESSMENT
UNIV. OF SOUTHERN MAINE
96 FALMOUTH ST
PO BOX 9300
PORTLAND, ME 04104-9300
207-780-4683

*8038
MARGARET PARK
STUDENT SERVICES ADVISOR
LEWISTON - AUBURN COLLEGE
51 WESTMINSTER ST
LEWISTON, ME 04240
207-753-6552

MARYLAND

9044
LIZ CLARY
EXTENDED LEARNING PROGRAM
ANNE ARUNDEL COMMUNITY COLLEGE
101 COLLEGE PKWY
ARNOLD, MD 21012
301-541-7305

9068
IDA BRANDON
CONTINUING EDUCATION
BOWIE STATE UNIVERSITY
JERICHO PARK RD
BOWIE, MD 20715
301-464-3000

*8065
DAVE KANNEY
ASSOC DIR OF ADMISSIONS
CAPITOL COLLEGE
11301 SPRINGFIELD RD
LAUREL, MD 20708
800-950-1992

9893
SANDY RAJASKI
REGISTRAR
CECIL COMMUNITY COLLEGE
1000 NORTHEAST RD
NORTH EAST, MD 21901
401-287-6060

*9555
GLORIA PETERSON
TEST CENTER ASSISTANTSUPERVISOR
CHARLES COUNTY COMMUNITY COLLEGE
MITCHELL RD, PO BOX 910
LA PLATA, MD 20646
301-934-7657

9872
LEE CAMPBELL
DEAN, CONTINUING EDUCATION
COLLEGE OF NOTRE DAME
4701 NORTH CHARLES ST
BALTIMORE, MD 21210
301-435-0100

*9508
DR. BETTY HOWARD
ACADEMIC SUPPORT PROGRAM
COLUMBIA UNION COLLEGE
WILKINSON HALL 441
7600 FLOWER AVE
TAKOMA PARK, MD 20912
301-891-4106

9032
CHERYL GRAHAM
CENTER FOR EDUCATIONAL SVCS
COMMUNITY COLLEGE OF BALTIMORE
2901 LIBERTY HEIGHTS AVE
BALTIMORE, MD 21215
301-396-0203

*8493
HUGO JACKSON
CONTINUING EDUC. COUNSELOR
COPPINS STATE COLLEGE
2500 WEST NORTH AVE
BALTIMORE, MD 21216
301-383-4500

9967
FRAN SMITHERS
DIRECTOR, COOPERATIVE
EDUCATION/JOB PLACEMENT
DUNDALK COMMUNITY COLLEGE
7200 SOLLERS POINT RD
BALTIMORE, MD 21222
401-285-9956

*9163
DANIEL E. BOCK
COUNSELOR
HAGERSTOWN JUNIOR COLLEGE
11400 ROBINWOOD DR
HAGERSTOWN, MD 21742
301-790-2800

*9164
VALERIE T. SWAIN
COORDINATOR OF TESTING
HARFORD COMMUNITY COLLEGE
401 THOMAS RUN RD
BEL AIR, MD 21014
410-836-4366

*8964
FREDERICA COFFEY
TEST CENTER SPECIALIST
HOWARD COMMUNITY COLLEGE
10901 LITTLE PATUXENT PKWY
COLUMBIA, MD 21044
410-772-4853

9557
CLARICE SOMERSALL
ASSESSMENT CENTER
MONTGOMERY COLLEGE
21 HUMANITIES BLDG 51
MANNAKEE ST
ROCKVILLE, MD 20850
301-251-7459

9282
JOHN ASH
TESTING SUPERVISOR
PRINCE GEORGE'S COMMUNITY COLLEGE
301 LARGO RD
LARGO, MD 20772
301-322-0147

9329
MICHAEL DUNN
ASSISTANT REGISTRAR
ST. MARY'S COLLEGE OF MARYLAND
ST. MARY'S CITY, MD 20686
301-862-0336

8666
REGINALD THOMAS
ASSOC DIRECTOR OF ADMISSIONS
UNIVERSITY OF BALTIMORE
1420 NORTH CHARLES ST
BALTIMORE, MD 21201
410-837-4796

9357
DIANE LAMPE
ASST. VICE PRESIDENT, STUDENT SVCS & ACADEMIC COUNSELING
UNIVERSITY OF MARYLAND UNIVERSITY COLLEGE
STUDENT & FACULTY SVCS BLDG.
UNIV. BLVD AT ADELPHI RD
COLLEGE PARK, MD 20742-1621
301-985-7288

9939
DR. PAUL BREWER
DIRECTOR, STUDENT SERVICES
UNIVERSITY OF MARYLAND
EUROPEAN DIVISION
UNIT 29216
APO AE 09102

9940
RICK HESS
DIRECTOR, STUDENT SERVICES
UNIVERSITY OF MARYLAND
ASIAN DIVISION, BOX 0100
APO AP 96328-0100

*9560
CHRISTINE DOERFLER
ASSISTANT ACADEMIC DEAN
WASHINGTON BIBLE COLLEGE
6511 PRINCESS GARDEN PKWY
LANHAM, MD 20706
301-552-1400

MASSACHUSETTS

*8531
LIZ RECKO-MORRISON
COORDINATOR, ASSMT & TESTING
BERKSHIRE COMMUNITY COLLEGE
1350 WEST ST
PITTSFIELD, MA 01201
413-499-4660

8467
NIKKI ENOS
REGISTRAR
BRADFORD COLLEGE
320 SOUTH MAIN ST
BRADFORD, MA 01830
508-372-7161

9565
WILLIAM IANNELLI
PROGRAM COORDINATOR
BUNKER HILL COMMUNITY COLLEGE
NEW RUTHERFORD AVE
BOSTON, MA 02129
617-241-8600

*8571
DANIEL NEWMAN
DIR, CTR OF LRNG & ASSMT
CAMBRIDGE COLLEGE
1000 MASSACHUSETTS AVE
CAMBRIDGE, MA 02138
617-868-1000 EXT 185

*9176
CLARE NIVEN - BLOWERS
DIRECTOR, ASSESSMENT CENTER
CAPE COD COMMUNITY COLLEGE
2240 IYANOUGH ROAD
WEST BARNSTABLE, MA 02668-1599
508-362-2131 EXT: 4454

9562
KATHY COUGHLIN
CONTINUING EDUCATION
CURRY COLLEGE
1071 BLUE HILL AVE
MILTON, MA 02186
617-333-2092

*8067
LEWANN MINA
TESTING ADMINISTRATOR
EASTERN NAZARENE COLLEGE LEAD PROGRAM
162-180 OLD COLONY AVENUE
QUINCY, MA 02171
617-774-6706

8328
CINDY O'CALLAGHAN
DIRECTOR, ADULT STUDIES PROGRAM
CENTER FOR ADULT STUDIES
EMMANUEL COLLEGE
400 THE FENWAY
BOSTON, MA 02115
617-735-9700

*9985
ELIZABETH MARTIN
REGISTRAR
FITCHBURG STATE COLLEGE
160 PEARL ST
FITCHBURG, MA 01420
978-655-3137

8499
JOSEPH W. McNABB
DEAN OF ACADEMIC AFFAIRS
LABOURE JUNIOR COLLEGE
2120 DORCHESTER AVE
BOSTON, MA 02124
617-296-8300

8727
STEVEN M. BROWN
DEAN PMBI PROGRAM
LESLEY COLLEGE
29 EVERETT ST
CAMBRIDGE, MA 02138
617-868-9600

*8497
MARGARET CAMPBELL
ACADEMIC ADVISOR
MASSACHUSETTS COLLEGE OF LIBERAL ARTS
375 CHURCH ST
NORTH ADAMS, MA 01247-4100
413-662-5400

8155
MICHAEL BEHNKE
DIRECTOR OF ADMISSIONS
MASSACHUSETTS INST. OF TECH.
77 MASSACHUSETTS AVE
CAMBRIDGE, MA 02139
617-253-1000

8539
EILEEN FAGEN
DIRECTOR RECORDS & REG.
MIDDLESEX COMMUNITY COLLEGE
33 KEARNEY SQUARE
LOWELL, MA 01852
508-656-3226

9238
PETER TANDY
ADVISING
MOUNT WACHUSETT COMMUNITY COLLEGE
444 GREEN ST
GARDNER, MA 01440
508-632-6600

*8609
JOHN L. MICHELSEN
COMMUNITY SCHOOL DIRECTOR
NANTUCKET HIGH SCHOOL
RM 126
10 SURF SIDE RD
NANTUCKETT ISLAND, MA 02554

9637
DR. BOGUSIA BUCKLIN
ASSO. DEAN,ACADEMIC AFAIRS
NEWBURY COLLEGE
129 FISHER AVE, BOX 425
BROOKLINE, MA 02445
617-730-7048

9566
PETER M. ENGH
REGISTRAR
NICHOLS COLLEGE
PO BOX 5000
DUDLEY, MA 01571-5000
508-943-1560

*9253
ELIZABETH H. FRUTIGER
TESTING COORDINATOR
NORTH SHORE COMMUNITY COLLEGE
300 BROAD ST, E 232
LYNN, MA 01901
781-477-2135

*9927
BOB BENNETT
TEST ADMINISTRATOR
NORTHEASTERN UNIVERSITY
COUNSELING CENTER
302 ELL BUILDING
360 HUNTINGTON AVE
BOSTON, MA 02115-9959
617-373-2142

8552
LISA KADER
ASSOC DIRECTOR OF ADMISSIONS
QUINCY COLLEGE
34 CODDINGTON ST
QUINCY, MA 02169
617-984-1703

8538
SUSAN GATELY
DIRECTOR, EXPERIENCE BASED ED
QUINSIGAMOND COMMUNITY COLLEGE
670 WEST BOYLESTON ST
WORCESTER, MA 01606
508-853-2300

8809
NANCY FINE
ASSOC DIRECTOR OF ADMISSIONS
SUFFOLK UNIVERSITY
8 ASHBURTON PLACE
BOSTON, MA 02108
617-573-8000

*8617
RONA MATISOFF
EDUCATIONAL THERAPIST
VETS ADMIN MEDICAL CTR
1400 VETS OF FOREIGN WARS PKWY
WEST ROXBURY, MA 02132
617-323-7700 EXT 6775

9974
DOUG KENYON, COUNSELOR
WESTERN NEW ENGLAND COLLEGE
HANSCOM
397 MAIN ST
WOBURN, MA 01801
617-274-0309

9390
JUDY CADDEN
DIVISION OF CONTINUING ED
WESTERN NEW ENGLAND COLLEGE
1215 WILBRAHAM RD
SPRINGFIELD, MA 01119-2684
800-325-1122, X1471

MICHIGAN

*9572
M. LLOYD ERICKSON, Ph.D.
DIR, COUNSELING & TESTING
ANDREWS UNIVERSITY
123 BELL HALL
BERRIEN SPRINGS, MI 49104
616-471-3470

*8997
DELORIS MULDREW
DIRECTOR, COUNSELING AND ASSESSMENT
BAKER COLLEGE
1050 WEST BRISTOL RD
FLINT, MI 48507
703-628-6094

*9084
DUANE GOUPELL
MANAGER
CLAS TESTING CENTER
CENTRAL MICHIGAN UNIVERSITY
MOUNT PLEASANT, MI 48859
517-774-3163

8782
VICTORIA HUBBELL
DIRECTOR OF TESTING
CONCORDIA COLLEGE
4090 GEDDES RD
ANN ARBOR, MI 48105
313-995-7300

8567
CHRIS WEBER
DIRECTOR OF ASSESSMENT
CORNERSTONE COLLEGE
1001 EAST BELTLINE NE
GRAND RAPIDS, MI 49505-5897
616-222-1431

*8413
ELLEN BROWN
ASSOC. DEAN - CONTG EDUC.
DAVENPORT COLLEGE
415 EAST FULTON ST
GRAND RAPIDS, MI 49503
616-732-1070

*9148
MICHELLE SALL
ASSISTANT REGISTRAR/ADVISOR
DAVENPORT COLLEGE - HOLLAND CAMPUS
643 SOUTH WAVERLY RD
HOLLAND, MI 49423
616-395-4719

*8296
DIANNE REY
REGISTAR
DAVENPORT COLLEGE
220 EAST KALAMAZOO ST
LANSING, MI 48933
800-686-1600

*9125
ELLEN B. WAXMAN
ASSESSMENT/TESTING COORD
DELTA COLLEGE
1961 DELTA ROAD, D-101
UNIVERSITY CENTER, MI 48710
517-686-9338

9575
CAROL FALWELL
DIRECTOR, ADMISSIONS/ADVISING
DETROIT COLLEGE OF BUSINESS
ADULT ACCELERATED CAREER EDUCATION PROGRAM
23400 MICHIGAN AVE, PLAZA 20
DEARBORN, MI 48124
313-562-8600

*9564
NANCY MCGEE
ASSISTANT DEAN
DETROIT COLLEGE OF BUSINESS
27500 DEQUINDRE RD
WARREN, MI 48092-5209
810-558-8700

9147
PENNY PAO
MGR, OFFICE OF ASSESSMENT
FERRIS STATE UNIVERSITY
410 OAK ST - ALU 101
BIG RAPIDS, MI 49307
616-592-3628

8665
BRUCE KOCHER
DEAN OF STUDENT SVCS
KALAMAZOO VALLEY COMMUNITY COLLEGE
6767 WEST "O" AVE
KALAMAZOO, MI 49009
616-372-5000

*8187
NANCY J. OBEY
MANAGER
TESTING & ASSESSMENT CENTER
KELLOGG COMMUNITY COLLEGE
450 NORTH AVENUE
BATTLE CREEK, MI 49017-3397
616-965-4136

*9578
DONALD DYER
DIRECTOR OF COUNSELING
KIRTLAND COMMUNITY COLLEGE
10775 NORTH SAINT HELEN RD
ROSCOMMON, MI 48653
517-275-5121

8956
LINDA STEINBERGER
ASST ADMISSIONS OFFICE
LAKE MICHIGAN COLLEGE
2755 EAST NAPIER AVE
BENTON HARBOR, MI 49022-1899
616-927-8107

9194
JOHN CANSFIELD
STUDENT DEVELOPMENT SVC
LANSING COMMUNITY COLLEGE
419 NORTH CAPITOL AVE
LANSING, MI 48901
517-483-1957

9981
KEVIN POLLOCK
ADMISSIONS OFFICE
LAWRENCE TECHNOLOGICAL UNIVERSITY
21000 WEST TEN MILE RD
SOUTHFIELD, MI 48075
810-204-4000

*9581
DR. MARY D. BROCKI
PRIOR LEARNING
MADONNA UNIVERSITY
36600 SCHOOLCRAFT RD
LIVONIA, MI 48150
313-432-5668

9235
PAUL C. SCHMIDT
REGISTRAR
MONROE COUNTY COMMUNITY COLLEGE
1555 SOUTH RAISINVILLE RD
MONROE, MI 48161-9047
313-242-7300

9585
JOHN BAMFIELD
ASSOC DEAN OF RECORDS & REG
MUSKEGON COMMUNITY COLLEGE
221 SOUTH QUARTERLINE RD
MUSKEGON, MI 49442
616-777-0364

*9658
KATHY SEDLACEK
TESTING COORDINATOR
NORTHWESTERN MICHIGAN COLLEGE
1701 E. FRONT STREET
TRAVERSE CITY, MI 49686-3061
616-922-1049

9260
CARL F. VANDER WOUDE
PROVOST
NORTHWOOD INSTITUTE
3225 COOK RD
MIDLAND, MI 48640
517-837-4411

9588
VICKI STEWART
ASSISTANT REGISTRAR
SAGINAW VALLEY STATE UNIVERSITY
7400 BAY RD
UNIVERSITY CENTER, MI 48710
517-790-4000

*8059
PAT BOISSEAU
LEARNING ASSISTANCE CTR.
SCHOOLCRAFT COLLEGE
18600 HAGGERTY RD
LIVONIA, MI 48152
313-462-4400 EXT 5303

9308
AUDREY W. PARKER
ACADEMIC ADVISING
SIENA HEIGHTS COLLEGE
1247 SIENA HEIGHTS DR
ADRIAN, MI 49221
517-263-0731

*9632
DEBORAH J. CARTER
DEAN, ACADEMIC AFFAIRS
SIENA HEIGHTS UNIVERSITY
2755 EAST NAPIER AVE
BENTON HARBOR, MI 49022-1899
616-927-6711

*9321
ROBERT S. CAMPBELL
ASSOCIATE DEAN
SOUTHWESTERN MICHIGAN COLLEGE
CHERRY GROVE RD
DOWAGIAC, MI 49047
616-782-5113

*8329
PAUL JOHNSON
STUDENT SERVICES COORDINATOR
SPRING ARBOR COLLEGE
800 LIVINGSTON BLVD
PO BOX 1766, AREC
GAYLORD, MI 49735
517-731-1177

*8330
PATTI DeAGOSTINO
SAC/TRAVERSE CITY CENTER
SPRING ARBOR COLLEGE
2200 DENDRINOS DR, SUITE 200E
TRAVERSE, MI 49684
517-750-1200

*8550
KATHE HEUMAN
GRADUATION AUDITOR
SPRING ARBOR COLLEGE
106 EAST MAIN ST
SPRING ARBOR, MI 49283
517-750-6518

8669
DON MYRIK
DEAN OF FACULTY
SUOMI COLLEGE
HANCOCK, MI 49930
906-482-5300

*8534
BONNIE NASKI
ASST. DIR., ACADEMIC ADVISING
UNIVERSITY OF DETROIT - MERCY
8200 WEST OUTER DRIVE
PO BOX 19900
DETROIT, MI 48219-0900
313-993-6015

9950
KEN SCHAEFER
COORDINATOR, TRANSFER RELATIONS & ARTICULATION
WESTERN MICHIGAN UNIVERSITY
OFFICE OF ADMISSIONS & ORIENTATION
KALAMAZOO, MI 49008-5120
616-387-2000

MINNESOTA

*9060
SHARON REBERS
COUNSELING CENTER
BEMIDJI STATE UNIVERSITY
1500 BIRCHMONT DR, NE
BEMIDJI, MN 56601
218-755-2000

*8783
KATHY OLSON
TESTING COORDINATOR
BETHEL COLLEGE
3900 BETHEL DR
ST.PAUL, MN 5[illegible]12-6999
612-635-8007

*8289
ROY ANDROLI
COUNSELOR
CENTRAL LAKES COLLEGE
501 WEST COLLEGE DR
BRAINERED, MN 56401
218-825-8016

*9110
CATHY JOHNSON
ADMISSIONS DATA ASSISTANT
COLLEGE OF ST. SCHOLASTICA
1200 KENWOOD
DULUTH, MN 55811
218-723-6464

8362
CHERIE R. HATLEM
REGISTRAR
CONCORDIA COLLEGE
901 SOUTH 8th ST
MOORHEAD, MN 56562
218-299-3250

9591
PAUL GEDDEN
REGISTRAR
CROWN COLLEGE
COUNTY RD 30 AND 92
ST. BONIFACIOUS, MN 55375
612-446-4100

8344
MARY D. ANDREA
LAKEWOOD COMMUNITY COLLEGE
3401 CENTURY AVE NORTH
WHITE BEAR LAKE, MN 55110

9258
DENNIS BENDICKSON
REGISTRAR
NORTHLAND COMMUNITY & TECH. COLLEGE
1101 HWY #1 EAST
THIEF RIVER FALLS, MN 56701
218-681-0858

*8060
TIFFANI WATTS
OFFICE ASSISTANT
FOCUS 15
NORTHWESTERN COLLEGE
3003 SNELLING AVE NORTH
ST. PAUL, MN 55113
612-631-5200

9994
BETTY STREHLOW
DIRECTOR
RIDGEWATER COLLEGE
PO BOX 1097
WILLMAR, MN 56201
320-235-5114

9593
SHERWOOD REID
ADMISSIONS
SAINT CLOUD STATE UNIVERSITY
SAINT CLOUD, MN 56301
612-255-0121

9980
MICHAEL E. GALKOWSKI
REGISTRAR
ST. MARY'S UNIVERSITY
700 TERRACE HEIGHTS, #37
WINONA, MN 55987
507-457-1428

*8787
GWENETH KAEBNICK
DEAN OF UNDERGRADUATE &
SPECIAL PROGRAMS
ST. MARY'S UNIV. OF MINNESOTA
2510 PARK AVE. SOUTH
MINNEAPOLIS, MN 55404
612-874-9877

*8006
DON ROBERTSON
COORD. COUNSELING/TESTING
SOUTHWEST STATE UNIVERSITY
BELLOWS 268
MARSHAL, MN 56258
507-537-7382

9596
LAUREL CARROLL
STUDENT SUPPORT SCS.
UNIVERSITY OF MINNESOTA
231 PILLSBURY DR SE
MINNEAPOLIS, MN 55455
612-625-2006

9988
JOHN IRVINE
DIRECTOR
UNIVERSITY OF MINNESOTA
MORRIS, MN 56267
612-589-6060

9393
JAMES MOOTZ
DIRECTOR OF ADMISSIONS
WINONA STATE UNIVERSITY
WINONA, MN 55987
507-457-5000

*8363
DON FLEMING
DEAN OF STUDENT SERVICES
WORTHINGTON COMMUNITY COLLEGE
1450 COLLEGE WAY
WORTHINGTON, MN 56187

MISSISSIPPI

8004
LAPLOSE JACKSON
COUNSELING & TESTING
ALCORN STATE UNIVERSITY
LORMAN, MS 39096
601-877-6100

**8243
BETTY LAMB
DIRECTOR, ADULT STUDENT SVCS
BELHAVEN COLLEGE
1500 PEACHTREE ST
JACKSON, MS 39202-1789
601-968-5926

9132
DAVID McCULLOOUGH
CONTINUING EDUCATION
EAST MISSISSIPPI JUNIOR COLLEGE
PO BOX 218
SCOOBA, MS 39358
601-476-8442

*8996
SHARON LEGGETT
TEST COORDINATOR
HINDS COMMUNITY COLLEGE
BOX 1296 HCC
RAYMOND, MS 39154
601-857-3219

8968
EVERETTE L. WITHERSPOON
VP, ACADEMIC AFFAIRS
JACKSON STATE UNIVERSITY
BOX 17199
JACKSON, MS 39217
601-968-2121

8988
DIANNE SPEED
DIRECTOR OF ADMISSIONS
JONES COUNTY JUNIOR COLLEGE
1000 COURT ST
ELLISVILLE, MS 39437
601-477-4025

8974
R. WAYNE PYLE
DIRECTOR OF ADMISSIONS
UNIVERSITY OF SOUTHERN MISSISSIPPI
BOX 504
HATTIESBURG, MS 39406
601-266-5000

MISSOURI

9085
DR. MICHAEL GRELLE
COORDINATOR, ASSESSMENT TESTING AND SERVICES
CENTRAL MISSOURI STATE UNIVERSITY
HUMPHREY BLDG - RM 223
WARRENSBURG, MO 64093
816-543-4919

9112
ALLYSON PRESLEY
DIRECTOR OF EVALUATIONS
COLUMBIA COLLEGE
1001 ROGERS ST
COLUMBIA, MO 65216
573-875-7674

*9537
VELDA KOSSUTH
COLUMBIA COLLEGE
PO BOX 1701
ROLLA, MO 654202
573-341-3350

*9119
SANDRA ELKIN
REGISTRAR
CULVER-STOCKTON COLLEGE
CANTON, MO 63435
314-288-5221

*8745
GEORGE PUHRMANN
DIRECTOR OF TESTING
DRURY COLLEGE
TRUMAN EDUCATION CENTER
FT. LEONARD WOOD, MO 65473
573-329-4400

*8984
LOREN L. SLEMP
COUNSELOR
EAST CENTRAL COLLEGE
HWY 50 EAST & PRAIRIE DELL RD
UNION, MO 63084
314-583-5193 EXT 2443

*8859
NANCY COX
OFFICE OF ADMINISTRATION
HANNIBAL/LA GRANDE COLLEGE
2800 PALMYRA RD
HANNIBAL, MO 63401
573-221-3675 EXT 276

*8882
LA VERNE ANDERSON
PRESIDENT
LAEL COLLEGE & GRADUATE SCHOOL
3721 ST. BRIDGET
ST. ANN, MO 63074
314-426-7000

9201
RICHARD PHELPS
REGISTRAR
LONGVIEW COMMUNITY COLLEGE
500 SW LONGVIEW RD
LEE'S SUMMIT, MO 64081
816-672-2000

9986
BETTY KUTILEK
REGISTRAR'S OFFICE
MARYVILLE COLLEGE
13550 CONWAY RD
SAINT LOUIS, MO 63141
313-576-9477

*8776
CHRISTIE RODGERS
ASSOC DIR OF ACADEMIC SVCS
NATIONAL-LOUIS UNIVERSITY
12412 POWERSCOURT DR - #LL20
ST.LOUIS, MO 63131
314-822-2110

*9254
NANCY ASHER
TESTING SERVICES
N.E. MISSOURI STATE UNIVERSITY
VIOLETE HALL 176
KIRKSVILLE, MO 63501
816-785-4140

*9274
DEBRA McARTHUR
COORDINATOR, ACADEMIC SUPPORT SERVICES
PARK COLLEGE
8700 NW RIVER PARK DR - #46
PARKVILLE, MO 64152-3795
816-741-2000 EXT 6332

9276
JACK BITZENBURG
DEAN OF CONTINUING EDUCATION
PENN VALLEY COMMUNITY COLLEGE
3201 SW TRAFFICWAY
KANSAS CITY, MO 64111
816-932-7600

This directory lists colleges & universities that award credit for passing scores on DSSTs.

*** INSTITUTIONS THAT ADMINISTER DSSTs**

**** Some schools may restrict testing to students admitted to or enrolled in the institution or offer a limited testing schedule.**

Students should contact the test center to determine current policies and schedules for DANTES testing.

9608
ROSI REISIG
REGISTRAR
ROCKHURST COLLEGE
1100 ROCKHURST RD
KANSAS CITY, MO 64110
816-501-4057

***9435**
JEWEL EGGLEY
DIRECTOR OF TESTING
TESTING SERVICES MS 3970
SOUTHEAST MISSOURI STATE UNIVERSITY
ONE UNIVERSITY PLAZA
CAPE GIRARDEAU, MO 63701
573-651-2836

***9539**
BRENT THOMAS
TESTING TECHNICIAN
COUNSELING & TESTING CENTER
SOUTHWEST MISSOURI STATE UNIVERSITY
901 SOUTH NATIONAL BOX 91
SPRINGFIELD, MO 65804
417-836-5116

***9609**
DON TANNER
DEAN OF CONTINUING EDUCATION
ST. LOUIS COMMUNITY COLLEGE
FLORISSANT VALLEY
3400 PERSHALL RD
ST. LOUIS, MO 63135
314-595-4444

9611
PAUL OWOC
EVENING DIVISION
ST. LOUIS UNIVERSITY
221 NORTH GRAND BLVD
SAINT LOUIS, MO 63103
314-658-2330

9613
JOYCE KIMBALL
DEAN OF STUDENTS
STATE FAIR COMMUNITY COLLEGE
3201 WEST 16th ST
SEDALIA, MO 65301
816-530-5800

***8838**
LESLIE KHRAIS
TESTING ADMINISTRATION SVCS
UNIVERSITY OF MISSOURI
220 PARKER HALL
COLUMBIA, MO 65211
573-882-4801

9358
WILSON BERRY
DIRECTOR OF ADMISSIONS
UNIVERSITY OF MISSOURI
5100 ROCKHILL RD
KANSAS CITY, MO 64110
816-276-1000

***8620**
WILBUR M. SIMON
DIRECTOR OF ASSESSMENT
UNIVERSITY OF MISSOURI - ST LOUIS
8001 NATURAL BRIDGE RD
ST. LOUIS, MO 63121-4499
314-516-5198

***9990**
ANNE HURST
DEPARTMENT ASSISTANT
WEBSTER UNIVERSITY
1740 WEST 92nd ST
KANSAS CITY, MO 64114

***9991**
ROBERT C. CHAMBERLIN
DIRECTOR OF ACADEMIC ADVISING
WEBSTER UNIVERSITY
470 EAST LOCKWOOD
ST. LOUIS, MO 63119
314-968-6972

MONTANA

***8566**
RONELDA MOGSTA
ADM ASST TO VICE PRES.
UNIVERSITY OF GREAT FALLS
1301 20th ST SOUTH
GREAT FALLS, MT 59405
406-791-5308

***8580**
DAVID R. MILLER
DEAN OF INSTRUCTION
FORT PECK COMMUNITY COLLEGE
US HWY 2 EAST
POPLAR, MT 59225
406-768-5551

8950
DALE E. OBERLANDER
DEAN OF STUDENT SERVICES
MILES COMMUNITY COLLEGE
2715 DICKINSON
MILES CITY, MT 59301
406-232-3031

***8585**
PAUL BEATTY
ASSOC DEAN OF STUDENTS
MONTANA TECH OF THE UNIV. OF MONTANA
1300 WEST PARK ST
BUTTE, MT 59701
406-496-4477

***8041**
AMY BEATTY
ASSISTANT DIRECTOR
DEGREE COMPLETION PRGM
ROCKY MOUNTAIN COLLEGE
1511 POLY DR
BILLINGS, MT 59102
406-657-1168

8342
BOB VAN GUNTEN
ADULT & CONTINUING EDUCATION
SALISH KOOTENAI COMMUNITY COLLEGE
PO BOX 117
PABLO, MT 59855
406-675-4800

***8129**
HELEN UTSOND
TEST SUPERVISOR
UNIVERSITY OF MONTANA
LODGE 148 CAMPUS DR
MISSOULA, MT 59812
406-243-0211

NEBRASKA

***9059**
INGEBORG JACOBS
ACADEMIC SUPPORT SVCS
BELLEVUE UNIVERSITY
1000 GALVIN RD SOUTH
BELLEVUE, NE 68005
800-756-7920, X3731

9305
DALE E. WILLIAMSON
REGISTRAR
CHADRON STATE COLLEGE
1000 MAIN STREET
CHADRON, NE 69337
308-432-6221
1-800-CHADRON

***8901**
JO CARPER
DIRECTOR, ACE/PONCI & SPECIAL PROGRAMS
CLARKSON COLLEGE
OMAHA, NE 68131-2739
402-559-3100

***8650**
MARY C. SCHLUETER
CO-DIRECTOR LEARNING/SUPPORT CENTER
COLLEGE OF SAINT MARY
1901 SOUTH 72nd ST
OMAHA, NE 68124
402-399-2634

***8765**
JERRIS McCOLLUM
STACY CREED
TEST ADMINISTRATORS
CONCORDIA COLLEGE
800 NORTH COLUMBIA AVE
SEWARD, NE 68434
402-643-7336

***8999**
NANCY PENNA
REGISTRAR
DANA COLLEGE
2848 COLLEGE DR
BLAIR, NE 68008
402-426-9000

***9620**
JOHN D. HOLMES
REGISTRAR
GRACE UNIVERSITY
1311 SOUTH 9th ST
OMAHA, NE 68108
402-449-2838

***8684**
JO KEELER
COUNSELING & TESTING
McCOOK COMMUNITY COLLEGE
McCOOK, NE 69001
308-345-6303

*8032
BARBARA WEGNER
EDUCATION SERVICES SPECIALIST
METROPOLITAN COMMUNITY COLLEGE
2732 EAST 23rd AVE
FREMONT, NE 68025
402-721-2507

8557
VALERIE SMITH
TESTING TECHNICIAN
METROPOLITAN COMMUNITY COLLEGE
FORT OMAHA CAMPUS
30th AND FORT ST
OMAHA, NE 68111
402-449-8400

8558
CHERYL GOODMAN
TESTING TECHNICIAN
METROPOLITAN COMMUNITY COLLEGE
2902 ED BABE GOMEZ AVE
OMAHA, NE 68111
402-738-4613

8556
LLOYD HUDGES
TESTING TECHNICIAN
METROPOLITAN COMMUNITY COLLEGE
ELKHOORN VALLEY CAMPUS
30th AND FORT ST
OMAHA, NE 68111
402-449-8400

*8831
ARDYCE GROSS
ASSOC DEAN OF STUDENTS
NORTHEAST COMMUNITY COLLEGE
801 EAST BENJAMIN AVE
NORFOLK, NE 68702
402-644-0413

8252
JESSE C. EDWARDS
ASSOCIATE DIRECTOR
PHYSICAN ASSISTANT PROGRAM
UNIVERSITY OF NEBRASKA
984300 NEBRASKA MEDICAL CENTER
OMAHA, NE 68198-4300
402-559-2741

8575
LINDA MURPHY
MEASUREMENTS
UNIVERSITY OF NEBRASKA
13 BANCROFT HALL
LINCOLN, NE 68588
402-472-7211

*9360
MARION FORTIN-WARA
TESTING CENTER
UNIVERSITY OF NEBRASKA
60th & DODGE STS
OMAHA, NE 68182
402-554-4800

8940
VIRGINIA FEELEY
COUNSELING CENTER
STUDENT CENTER
WAYNE STATE COLLEGE
1111 MAIN ST
WAYNE, NE 68787
402-375-7321

NEVADA

*8878
ROY BURROWS
COORDINATOR, ASSESSMENT CTR.
COMMUNITY COLLEGE OF SOUTHERN NEVADA
3200 EAST CHEYENNE AVE
NORTH LAS VEGAS, NV 89030
702-651-4050

9527
RICHARD EYDE
CENTER DIRECTOR
EMBRY-RIDDLE AERONAUTICAL UNIVERSITY - EXTENDED CAMPUS
NAS FALLON RESIDENT CENTER
PO BOX 623
FALLON, NV 89407-0623
702-423-4018

9361
JOHN WITTER
ASSOC DIRECTOR, ADMISSIONS
UNIVERSITY OF NEVADA
4505 SOUTH MARYLAND PKWY
LAS VEGAS, NV 89154
702-739-3371

NEW HAMPSHIRE

*8478
TERESA McDONNELL
ASSOCIATE DEAN
COLLEGE FOR LIFELONG LEARNING
125 NORTH STATE ST
CONCORD, NH 03301-6430

*8749
LORRAINE FARRAR
COORDINATOR LEARNING SVCS
COLLEGE FOR LIFELONG LEARNING.
PO BOX 385
CONWAY, NH 03818
603-447-3970

*8803
LORRAINE FARRAR
COORDINATOR OF ADM. SERVICES
COLLEGE FOR LIFELONG LEARNING.
148 PLEASANT ST
BERLIN, NH 03570
603-752-4224

*8711
PAULINE ANGIER
EDUATIONAL ASISTANT
COLLEGE FOR LIFELONG LEARNING
325 MT. SUPPORT RD, SUITE 2
LEBANON, NH 03766
603-448-6797

*8263
ROBERTA POLSHUK
PROGRAM COORDINATOR
COLLEGE FOR LIFELONG LEARNING
507B SOUTH ST
BOW, NH 03304
603-226-9000

9179
HAZEL A. SCHELPER
REGISTRAR
FRANKLIN PIERCE COLLEGE
PO BOX 60, COLLEGE RD
RINDGE, NH 03461-0060
603-899-4097

9966
ELAINE MINNEHAN-CARON
ASSOCIAT DEAN
HESSER COLLEGE
3 SUNDIAL AVE
MANCHESTER, NH 03103
603-668-6660

8945
DOROTHY JOHNSEN
SR. V.P., ENROLLMENT MGMT
McINTOSH COLLEGE
23 CATARACT AVE
DOVER, NH 03820
603-742-1234

9245
NANCY DAVIS
ASSISTANT REGISTRAR
NEW HAMPSHIRE COLLEGE
2500 NORTH RIVER RD
MANCHESTER, NH 03106
603-668-2211 EXT 2125

9017
MARICA MERRICK
ACADEMIC AFFAIRS
NEW HAMPSHIRE TECHNICAL COLLEGE
277 R. PORTSMOUTH AVE
STRATHAM, NH 03885
603-772-1194

*8673
ART LeBLANC
ENGINEERING/TECHNOLOGY DEPARTMENT
NEW HAMPSHIRE TECHNICAL INSTITUTE
11 INSTITUTE DR
CONCORD, NH 03301
603-225-1800

*9863
DEBORAH J. TOBINE
EDUCATIONAL ASSISTANT
PLYMOUTH STATE COLLEGE
SPEARE ADMIN.BLDG , RM 126
PLYMOUTH, NH 03264
603-535-2685

9441
BETH BEYERLE
ASSISTANT DEAN, UNDERGRADUATE EVENING SCHOOL
RIVIER COLLEGE
420 MAIN STREET
NASHUA, NH 03060-5086
603-888-1311 EXT 8229

NEW JERSEY

8509
JOHN T. MAY
DEAN OF ACADEMICS
ATLANTIC COMMUNITY COLLEGE
MAYS LANDING, NJ 08330
609-343-5099

***9463**
JANET MASIELLO
TEST ADMINISTRATOR
BERGEN COMMUNITY COLLEGE
OFFICE OF TESTING & ASSESSMENT,
ROOM L-121
400 PARAMUS ROAD
PARAMUS, NJ 07652-1595
201-447-7876

***9071**
LAURA MICELI
COORDINATOR, TESTING SERVICES
BROOKDALE COMMUNITY COLLEGE
765 NEWMAN SPRINGS RD
LINCROFT, NJ 07738
908-842-1900

***9138**
LISA DIBISCEGLIE
DIRECTOR, CENTER FOR CONTINUING EDUCATION
CALDWELL COLLEGE
9 RYERSON AVE
CALDWELL, NJ 07006
973-618-3280

***8391**
TASHA JONES
TEST ADMINISTRATOR
CAMDEN COUNTY COLLEGE
BOX 200
BLACKWOOD, NJ 08012
609-227-7200, X4710

***9447**
ROBERT QUADE
DIRECTOR OF LIFE LEARNING
CENTENARY COLLEGE
400 JEFFERSON STREET
HACKETTSTOWN, NJ 07840

9638
ALFRED BRIDGES
DEAN ADMISSIONS & RECORDS
COLLEGE OF NEW JERSEY
HILLWOOD LAKES CN 4700
TRENTON, NJ 08650
609-771-1855

***8282**
SISTER DOROTHY LAZARICK
DIRECTOR, EVENING DIVISION
GEORGIAN COURT COLLEGE
900 LAKEWOOD AVE
LAKEWOOD, NJ 08701
732-364-8218

***8175**
MOLLY PENNELL, DIRECTOR
PROFESSIONAL & CONTINUING ED
GLOUCESTER COUNTY COLLEGE
HUMAN RESOURCE & ECONOMIC DEVELOPMENT CENTER
520 HERRON DR, SUITE 113
PURELAND INDUSTRIAL PARK
BRIDGEPORT, NJ 08014

***8355**
DR. PERPETUA RAZON
DIRECTOR ASSMT & TESTING
HUDSON COUNTY COMMUNITY COLLEGE
168 SIP AVE
JERSEY CITY, NJ 07306
201-714-2166

8427
JOHN C. DUFF
ASST. DIRECTOR OF ACADEMIC CAREER PLANNING
JERSEY CITY UNIVERSITY
2039 KENNEDY BLVD
JERSEY CITY, NJ 07305
201-200-3005

***9220**
MARILYN PENDER
ASST TO VP FOR ACADEMIC & STUDENT AFFAIRS
MERCER COUNTY COMMUNITY COLLEGE
1200 OLD TRENTON RD
TRENTON, NJ 08690
609-586-4800 EXT 3298

***9932**
NANCY BRUNO
DIRECTOR OF TESTING
MIDDLESEX COUNTY COLLEGE
155 MILL RD
EDISON, NJ 08818
908-906-2508

9234
ELIZABETH A. MARTIN
DIRECTOR OF ADULT ADMISSIONS
MONMOUTH UNIVERSITY
CEDAR & NORWOOD AVE
WEST LONG BRANCH, NJ 07764
908-571-3433

9635
RICHARD TAUBOLD
ADULT EDUCATION
MONTCLAIR STATE COLLEGE
UPPER MONTCLAIR, NJ 07043
201-893-4000

***8512**
FRED HARRIS
COUNSELING CENTER
NEW JERSEY INSTITUTE OF TECHNOLOGY
UNIVERSITY HEIGHTS
NEWARK, NJ 07102
201-596-3000

***9956**
JANET NELSON
COORDINATOR OF ARTICULATED & PROGRAMS TESTING
OCEAN COUNTY COLLEGE
COLLEGE DR, PO BOX 2001
TOMS RIVER, NJ 08754
908-255-0496

***9011**
LANA BROWNE
DIRECTOR OF COUNSELING
RARITAN VALLEY COMMUNITY COLLEGE
PO BOX 3300
SOMERVILLE, NJ 08876
908-526-1200 EXT 8331

8465
MARY PINNEY
ASSOCIATE DEAN
RIDER UNIVERSITY
2083 LAWRENCEVILLE RD
LAWRENCEVILLE, NJ 08691
609-896-5033

***8042**
Dr. CLAUDE D'AMICO
DIRECTOR OF TESTING
ROWAN UNIVERSITY
201 MULLICA HILL RD
HAWTHORNE HALL
GLASSBORO, NJ 08028
609-256-4238

8818
JULIA BRANAN
ASSOCIATE DEAN OF ACADEMIC AFFAIRS
SALEM COMMUNITY COLLEGE
460 HOLLYWOOD AVE
CARNEYS POINT, NJ 08069-2799
609-351-2641

***8712**
JOE MALONE
DIRECTOR
SOMERSET TECHNOLOGICAL VOCATIONAL INSTITUTE
NORTH BRIDGE ST
PO BOX 6350
BRIDGEWATER, NJ 08807
908-526-8900

8324
WILLIAM CUNNINGHAM
DEPARTMENT OF VOCATIONAL EDUCATION STUDIES
SOUTHERN ILLINOIS UNIVERSITY
BLDG 1911, RM 9
McGUIRE AFB, NJ 08641

***8177**
ELEANOR HANSEN
TEST PROCTOR
STOCKTON STATE COLLEGE
JIM LEEDS RD, D-108
POMONA, NJ 08240

***9001**
MAUREEN WOODRUFF
OFFICE OF TEST ADMIN.
THOMAS EDISON STATE COLLEGE
101 WEST STATE ST
TRENTON, NJ 08608-1176
609-984-1181

9345
ADMISSIONS OFFICE
UNION COUNTY COLLEGE
1033 SPRINGFIELD AVE
CRANFORD, NJ 07016
908-709-7596

This directory lists colleges & universities that award credit for passing scores on DSSTs.

*** INSTITUTIONS THAT ADMINISTER DSSTs**

**** Some schools may restrict testing to students admitted to or enrolled in the institution or offer a limited testing schedule.**

Students should contact the test center to determine current policies and schedules for DANTES testing.

*8542
MAGGIE VAN BENSCHOTEN
TEST ADMINISTRATOR
WARREN COUNTY COMMUNITY COLLEGE
475 ROUTE 57 WEST
WASHINGTON, NJ 07882-9605
908-835-2354

*9639
RACHEL ANZALDO
DIRECTOR, TESTS AND MEASUREMENTS
WILLIAM PATERSON COLLEGE
HUNZIKER WING 220
300 POMPTON RD
WAYNE, NJ 07470
973-720-3107

NEW MEXICO

8967
IDA STOVER
DIRECTOR, ADMISSIONS AND RECORDS
EASTERN NEW MEXICO UNIVERSITY
PO BOX 6000
ROSWELL, NM 88201
505-624-7145

8138
LOUISE CHAMBERLIN
DIRECTOR OF ADMISSIONS
NEW MEXICO INSTITUTE OF MINING AND TECHNOLOGY
SOCORRO, NM 87801
505-835-5011

*8703
MIKE COSTELLO
DEAN OF STUDENTS
NORTHERN NEW MEXICO COMMUNITY COLLEGE
1002 NORTH ONATE ST
ESPANOLA, NM 87532
505-747-2120

8332
BARBRA TUCCI
DIRECTOR OF ACADEMIC RECORDS
SANTA FE COMMUNITY COLLEGE
6401 RICHARDS AVE
SANTA FE, NM 87502
505-438-1264

9643
DAVID SANCHEZ
TESTING CENTER
UNIVERSITY OF NEW MEXICO
UNIVERSITY COLLEGE BUILDING RM 2
ALBUQUERQUE, NM 87131-1461
505-277-5345

NEW YORK

*8981
JEANNE CHARPENTIER
DIRECTOR OF ACADEMIC INFORMATION SERVICES
ADIRONDACK COMMUNITY COLLEGE
BAY RD
QUEENSBURY, NY 12804
518-743-2280

*8708
DR. WARD DEUTSCHMAN
DEAN OF ACADEMICS
BRIARCLIFFE COLLEGE
1055 STEWART AVE.
BETHPAGE, NY 11714
516-470-6000 EXT 522

*8513
CARMEN DEVITA
ACTING CHAIR, LEARNING ASSISTANCE CENTER
BROOME COMMUNITY COLLEGE
PO BOX 1017
BINGHAMTON, NY 13902
607-778-5038

8611
LEE HERMAN
COORDINATOR
CAYUGA COMMUNITY COLLEGE
FRANKLIN ST
AUBURN, NY 13021
315-255-2794

8136
BARRY KANE
REGISTRAR
COLGATE UNIVERSITY
13 OAK DR
HAMILTON, NY 13346
315-824-1000

*9473
MARIA KOLODZIEJ
ASSISTANT DIRECTOR OF ADMISSIONS
COLUMBIA GREENE COMMUNITY COLLEGE
4400 ROUTE 23
HUDSON, NY 12534
518-828-4181

*9677
ESTELLA E. LOUD
ASSESSMENT COORDINATOR
CORNING COMMUNITY COLLEGE
1 ACADEMIC DR
CORNING, NY 14830
607-962-9329

9128
SR. M. CARMEL Mc ENEANY
TRANSFER COUNSELOR
DOMINICAN COLLEGE
470 WESTERN HWY
ORANGEBURG, NY 10962
914-359-7800 EXT 274

8659
KATHLEEN B. ROWE
DIRECTOR, ENROLLMENT SERVICES
DOWLING COLLEGE
IDLE HOUR BLVD
OAKDALE, NY 11769
516-244-3030

9130
CARL MARCHESE
ASSOCIATE REGISTRAR
DUTCHESS COMMUNITY COLLEGE
53 PENDELL RD
POUGHKEEPSIE, NY 12601
914-471-4500

9563
BETH FARR
REGISTRAR'S ASSISTANT
ELMIRA COLLEGE
PARK PLACE
ELMIRA, NY 14901
607-735-1895

8373
KRISTI ROBERTS
CENTER DIRECTOR
EMBRY-RIDDLE AERO UNIVERSITY
KATTERBACH EDUCATION CENTER
APO, NY 09250
011-49-9802-8757

8551
B. PAUL HODAN
DIRECTOR STUDENT SVCS
ERIE COMMUNITY COLLEGE
6205 MAIN ST
WILLIAMSVILLE, NY 14221
716-851-1588

*9676
DANIEL LS. HAYNES
DSS PROJECT COORD
CENTER FOR AVISEMENT/CAREER SVCS
FINGER LAKES COMMUNITY COLLEGE
4355 LAKE SHORE DR
CANANDAIGUA, NY 14424
716-394-3500 EXT 460

*8636
JOHN HOUSTON
ASSOCIATE DEAN OF IGNATIUS COLLEGE
FORDHAM UNIVERSITY
441 FORDHAM RD
BRONX, NY 10458
718-817-4600

9679
JOY GREEN
TECHNICAL ASSISTANT
GENESEE COMMUNITY COLLEGE
ONE COLLEGE RD
BATAVIA, NY 14020
716-343-0055

9166
NANCY HULSE
REGISTRAR
HARTWICK COLLEGE
BRESEE HALL
ONEONTA, NY 13820
607-431-4200

This directory lists colleges & universities that award credit for passing scores on DSSTs.

*** INSTITUTIONS THAT ADMINISTER DSSTs**
**** Some schools may restrict testing to students admitted to or enrolled in the institution or offer a limited testing schedule.**

Students should contact the test center to determine current policies and schedules for DANTES testing.

*8789
DR. WILLIS BEARDSLEY
DIRECTOR OF ASSESSMENT
HOUGHTON COLLEGE
ONE WILLARD AVE
HOUGHTON, NY 14744
716-567-9456

*9680
SUSAN CARVIN
COUNSELOR
HUDSON VALLEY COMMUNITY COLLEGE
80 VANDENBURGH AVE
TROY, NY 12180
518-270-7320

9180
JOHN G. PHILLIPS
ASSOCIATE DEAN
JEFFERSON COMMUNITY COLLEGE
COFTEEN ST
WATERTOWN, NY 13601
315-786-2249

8027
NORBERT HENRY
DIRECTOR OF CONTINUING ED
LE MOYNE COLLEGE
SYRACUSE, NY 13124
315-445-4100

*9187
RABBI SHOLOM G. GINZBERG, DEAN
MA'A LOT - NEW YORK
A DIVISION OF THE ZAIDNER INSTITUTE
1066 EAST 23RD STREET
BROOKLYN, NY 11210-3638
718-377-0222

8686
SR. NANCY QUINLAN
REGISTRAR
MARIA COLLEGE OF ALBANY
700 NEW SCOTLAND AVE
ALBANY, NY 12208
518-438-3111, X224

9215
ELLEN CAMPBELL
CONTINUING EDUCATION OFFICE
MARYMOUNT COLLEGE
TARRYTOWN, NY 10591
914-631-3200

8664
RAYMOND JOHNSON
ACADEMIC DEAN
MATER DEI COLLEGE
RIVERSIDE DR
OGDENSBURG, NY 13669
315-393-5930

*9031
JENIFER BAVIFORD
DIRECTOR, OFF CAMPUS & SPECIAL PROGRAMS
MEDAILLE COLLEGE
18 AGASSIZ CIRCLE
BUFFALO, NY 14214
716-884-3281, X235

9221
EUGENE WHEELER
ADMISSIONS COUNSELOR
MERCY COLLEGE
555 BROADWAY
DOBBS FERRY, NY 10522
914-693-7600

9232
PAMELA NIGRO
ACADEMIC ADVISING
MOHAWK VALLEY COMMUNITY COLLEGE
1101 SHERMAN DR
UTICA, NY 13501
315-792-5301

*9156
ALFRED J. POLIZZI
TEST ADMINISTRATOR
ADMISSIONS COUNSELOR
MONROE COMMUNITY COLLEGE
1000 EAST HENRIETTA ROAD
ROCHESTER, NY 14623
716-292-3860

*9652
GERALD JILBERT
REGISTRAR
MOUNT SAINT MARY COLLEGE
330 POWELL AVE
NEWBURGH, NY 12550
914-569-3258

9653
ELAINE A. MISSEL
TRANSFER CREDIT EVALUATOR
NAZARETH COLLEGE OF ROCHESTER
4245 EAST AVE
ROCHESTER, NY 14618-3790
716-389-2873

9975
ELIZABETH ROSS
REGISTRAR
NEW SCHOOL FOR SOCIAL RESEARCH
66 WEST 12TH ST
NEW YORK, NY 10011
212-229-5650

*9248
EILEEN ZABEL
ADMINISTRATOR
NEW YORK INSTITUTE OF TECHNOLOGY
PO BOX 9029
CENTRAL ISLIP, NY 11722-9029
516-348-3312

8677
ELINOR AZENBERG
DIRECTOR RE-ENTRY PROGRAMS
ADULT DEGREE STUDIES
NEW YORK UNIVERSITY
SHIMKIN HALL
NEW YORK, NY 10003
813-638-1426

*9249
ELIZABETH FULWELL
ASSOC DEAN/ACADEMIC SVCS
NIAGARA COUNTY COMMUNITY COLLEGE
3111 SAUNDERS SETTLEMENT RD
SANBORN, NY 14132
716-731-3271

*8965
ELENA MURPHY
DIRECTOR OF ASSESSMENT
NYACK COLLEGE
ADULT DEGREE COMPLETION PROGRAM
ONE SOUTH BLVD
NYACK, NY 10960
914-358-5360

9266
SANDRA F. MARK
ACADEMIC AFFAIRS
ORANGE COUNTY COMMUNITY COLLEGE
115 SOUTH ST
MIDDLETOWN, NY 10940
914-341-4768

*8800
DORIS BUCK
COORD OF ADULT COUNSELING
ORANGE-ULSTER BOCES
GIBSON RD
GOSHEN, NY 10924
914-294-2295

*8793
PROF. JOSEPH V. DUMBRA
DIRECTOR, ADULT SERVICES
PACE UNIVERSITY
MIDTOWN CENTER
551 FIFTH AVE - 8th FL.
NEW YORK, NY 10176
212-346-1700

8561
HELEN BEUKA
TRANSFER COORDINATOR
PACE UNIVERSITY
PLEASANTVILLE, NY 10570
914-773-3746

9370
ANDREA LUCAS
DIRECTOR OF TESTING
REGENTS COLLEGE
7 COLUMBIA CIRCLE
ALBANY, NY 12065
518-464-8592

*8961
LINDA GARFINKEL
TESTING ADMINISTRATOR
COUNSELING CENTER
ROCHESTER INSTITUTE OF TECHNOLOGY
114 LOMB MEMORIAL DRIVE
ROCHESTER, NY 14623-5608
716-475-2261

8530
VICKI ASPRIDY
ASSISTANT REGISTRAR
ROCHESTER UNIVERSITY
27 LOMB MEMORIAL DR
ROCHESTER, NY 14623-5603

*8719
ROBERT GLEASON
DIRECTOR, LIBRARY SVCS
ROCKLAND COMMUNITY COLLEGE
145 COLLEGE ROAD
SUFFERN, NY 10901
914-574-4472

This directory lists colleges & universities that award credit for passing scores on DSSTs.

*** INSTITUTIONS THAT ADMINISTER DSSTs**
**** Some schools may restrict testing to students admitted to or enrolled in the institution or offer a limited testing schedule.**

Students should contact the test center to determine current policies and schedules for DANTES testing.

9311
ROBERT VAN METER
DIRECTOR
SKIDMORE COLLEGE
SARATOGA SPRINGS, NY 12866
518-584-5000

9327
DENNIS CROWLEY
REGISTRAR
ST. JOHN FISHER COLLEGE
3690 EAST AVE
ROCHESTER, NY 14618
716-385-8015

*9659
DR. W. DAVID HARMON
ASSOC DIR, COUNSELING CENTER
ST. JOHN'S UNIVERSITY
MARILLAC HALL, RM 130
8000 UTOPIA PKWY
JAMAICA, NY 11439
718-990-6383

*8894
CYNTHIA ANDREI
ASSISTANT FOR STUDENT SERVICES
DIVISION OF GENERAL STUDIES
ST. JOSEPH'S COLLEGE
265 CLINTON AVE
BROOKLYN, NY 11205-3688
718-399-0068

*9886
MARIE LOSQUADRO
ASSISTANT TO DEAN
ST. JOSEPH'S COLLEGE
155 WEST ROE BLVD
PATCHOGUE, NY 11772-2399
516-447-3261

9332
MILLIE ALEXIOU
REGISTRAR
ST. THOMAS AQUINAS COLLEGE
RT 340
SPARKILL, NY 10976
914-359-9500

8354
DOUGLAS STEELE
DIRECTOR OF ADMISSIONS
SUFFOLK COUNTY COMMUNITY COLLEGE
533 COLLEGE RD
SELDEN, NY 11784
516-451-4110

9662
JAMES SULLIVAN
TRANSFER COORDINATOR
SUNY AT BINGHAMTON
PO BOX 600
ACADEMIC ADVISING OFFICE
BINGHAMTON, NY 13902
607-777-6305

9323
OFFICE OF ACADEMIC ADVISEMENT
SUNY AT BROCKPORT
350 NEW CAMPUS DR
BROCKPORT, NY 14420
716-395-2711

8657
DR. DALE E. BOWER
VP FOR ACADEMIC AFFAIRS
SUNY COLLEGE OF TECH/CANTON
CORNELL DR
CANTON, NY 13617
315-386-7202

*8050
MELINDA VAN ETTEN
COUNSELING CTR.
SUNY AT CORTLAND
B-44 VAN HOESEN HALL
P.O. BOX 2000
CORTLAND, NY 13045
607-753-4728

**8737
THOMAS F. McELROY
COORD., STUDENT SVCS & ASSMT
SUNY - EMPIRE STATE COLLEGE
GENESSEE VALLEY CENTER
8 PRINCE ST
ROCHESTER, NY 14607-1406
716-244-3641

**8739
MARGARET EVANS
COORD., ASSMT & LRNG RESRCS
SUNY - EMPIRE STATE COLLEGE
METROPOLITAN CENTER
225 VARICK T
NEW YORK, NY 10014-4382
212-647-7800

**8740
SHARON PERRY
TEST ADMINISTRATOR
ASSESSMENT & LEARNING RESCRS
SUNY - EMPIRE STATE COLLEGE
CENTRAL NEW YORK CENTER
219 WALTON ST
SYRACUSE, NY 13202-1226
315-472-5799

**8732
JOSEPHINE LAU
DIRECTOR, PROGRAM REVIEW/ASSESSMENT
SUNY - EMPIRE STATE COLLEGE
2 UNION AVE
SARATOGA SPRINGS, NY 12866
518-587-2100, X211

**8733
MARY DANN
COORDINATOR OF ASSESSMENT & LEARNING RESOURCES
SUNY - EMPIRE STATE COLLEGE
NORTHEAST CENTER
845 CENTRAL AVE
ALBANY, NY 12206-1506
518-485-5964

**8736
EMILY RILEY
TEST ADMINISTRATOR
SUNY - EMPIRE STATE COLLEGE
617 MAIN ST
MARKET ARCADE BLDG
BUFFALO, NY 14203-1498
716-853-7700

**8738
LESLIE ELLIS
COORDINATOR OF LEARNING RESOURCES AND ASSESSMENT
SUNY - EMPIRE STATE COLLEGE
HUDSON VALLEY CENTER
200 NORTH CENTRAL AVE
HARTSDALE, NY 10530-1999
914-948-6206

**8735
ED SAUERACKER
ASST DEAN FOR ASSESSMENT
SUNY - EMPIRE STATE COLLEGE
LONG ISLAND CENTER
TRAINOR HOUSE
223 STORE HILL RD
PO BOX 130
OLD WESTBURY, NY 11568-0130
516-997-4700

**9590
LISA BERRY
TEST ADMINISTRATOR
SUNY-EMPIRE STATE COLLEGE
AUBURN UNIT
CAYUGA COMMUNITY COLLEGE
197 FRANKLIN STREET
AUBURN, NY 13021-3099
315-255-2794

**9592
JANET SMITH
TEST ADMINISTRATOR
SUNY-EMPIRE STATE COLLEGE
BINGHAMTON UNIT
44 HAWLEY ST, 5TH FL.
BINGHAMTON, NY 13901-4451
607-721-8651

**9594
BETTY SLYE
TEST ADMINISTRATOR
SUNY-EMPIRE STATE COLLEGE
FORT DRUM UNIT
BLD T-21, ARMY LEARNING CTR
P.O. BOX 9
BLACK RIVER, NY 13612-0009
315-773-6139

**9595
DIANE WALKER
TEST ADMINISTRATOR
SUNY-EMPIRE STATE COLLEGE
THE ROTHSCHILD BUILDING
215 EAST STATE STREET
ITHACA, NY 14850-5549
607-273-4536

**9597
CONNIE WELLS
TEST ADMINISTRATOR
SUNY-EMPIRE STATE COLLEGE
PLATTSBURGH UNIT
101 BORAD ST, RM 403 SIBLEY HALL
PLATTSBURGH, NY 12901-2581
518-564-2837

**9599
DARLINE WISNIEWSKI
TEST AMINISTRATOR
SUNY-EMPIRE STATE COLLEGE
207 GENESSEE ST, SUITE 606
STATE OFFICE BULDING
UTICA, NY 13501-2812
315-793-2684

**9601
MAUREEN KRAVECK
COORDINATOR
SUNY EMPIRE STATE COLLEGE
317 WASHINGTON ST
4TH FLOOR, STATE OFFICE BLDG
WATERTOWN, NY 13601-3744
315-785-2418

**9602
CAROLE SOUTHWOOD
COORDINATOR
SUNY-EMPIRE STATE COLLEGE
LOCKPORT SENIOR CITIZEN CTR
9 ONTARIO ST
LOCKPORT, NY 14094-2815
716-434-0272

**9604
MARY ELLEN SHAUGHNESSY
TEST ADMINISTRATOR
SUNY-EMPIRE STATE COLLEGE
CATTARAUGUS COUNTY CAMPUS
244-250 NORTH UNION ST
OLEAN, NY 14760-2662
716-372-1661, X557

**9606
RAE ROHFELD
TEST ADMINISTRATOR
SUNY-EMPIRE STATE COLLEGE
CENTER FOR DISTANCE LEARNING
3 UNION AVE
SARATOGA SPRINGS, NY 12866-4391
518-587-2100, X300

**9610
CATHY PLUMMER
TEST ADMINISTRATOR
SUNY-EMPIRE STATE COLLEGE
4355 LAKESHORE DRIVE
CANANDAIGUA, NY 14424-8395
716-394-1110

**9614
CATHERINE MCALLITER
COORDINATOR
SUNY-EMPIRE STATE COLLEGE
STATE UNIV COLL OF FREDONIA
3173 MASON HALL
FREDONIA, NY 14063-1198
716-673-3575

**9616
KATHIE MOSHER
TEST ADMINISTRATOR
SUNY-EMPIRE STATE COLLEGE
22 WEST THIRD ST
CORNING, NY 14830-3198
607-962-1421

**9617
LESLIE ELLIS
TEST ADMINISTRATOR
SUNY-EMPIRE STATE COLLEGE
HIGHER EDUCATION CONSORTIUM
90 NORTH STREET
MIDDLETOWN, NY 10940
914-341-4110 or 914-948-6206

**9618
LESLIE ELLIS
TEST ADMINISTRATOR
SUNY-EMPIRE STATE COLLEGE
HERITAGE SQUARE
602 ROUTE 299
HIGHLAND, NY 12528-2848
914-883-5358 or 914-948-6206

**9619
LESLIE ELLIS
TEST ADMINISTRATOR
SUNY-EMPIRE STATE COLLEGE
240 NORTH MAIN ST
NEW CITY, NY 10956-5302
914-708-7010 or 914-948-6206

*8160
WAYNE CLAVERING
TEST ADMINISTRATOR
SUNY-FARMINGDALE
ROUTE 110
FARMINGDALE, NY 11735
516-420-2594

9324
ROGER B. SULLIVAN
DIRECTOR OF ADMISSIONS
SUNY INSTITUTE OF TECHNOLOGY
UTICA/ROME MARCY CAMPUS
PO BOX 3050
UTICA, NY 13504
315-792-7208

9669
LINDA SYRELL
DEAN, CONTINUING EDUCATION
SUNY AT OSWEGO
OSWEGO, NY 13126
315-341-2270

*9325
BETTE BROHEL
COORD, ADULT LEARNING SVCS
SUNY AT PLATTSBURGH
KEHOE 412
PLATTSBURGH, NY 12901
518-564-2050

8649
SUZANNE MANFRED
TRANSFER ARTICULATION
SUNY AT POTSDAM
DUNN HALL, RM 102
POTSDAM, NY 13676
315-267-2000

9114
DENNIS STRATTON
DIRECTOR OF ADMISSIONS
SUNY AT SYRACUSE
1 FORESTRY DR
SYRACUSE, NY 13210
800-777-7373

*8872
JANE HAMMOND
REGISTRAR & DIRECTOR
OF INSTITUTIONAL RESEARCH
TOMPKINS CORTLAND COMMUNITY COLLEGE
170 NORTH ST
DRYDEN, NY 13053
607-844-8211

*8674
SAM KARGBO
ASSOC DIRECTOR, EDUCATIONAL
OPPORTUNIITIES
UNIVERSITY OF BUFFALO
EDUCATIONAL OPPORTUNITY
CENTER
465 WASHINGTON ST
BUFFALO, NY 14203
716-849-6737

9380
ARLENE SULLIVAN
DIR., LEARNING SUPPORT CTR
VILLA MARIA COLLEGE
240 PINE RIDGE RD
BUFFALO, NY 14225
716-896-0700

NORTH CAROLINA

*9706
SUSAN GREENE
ASST DIRECTOR OF TESTING
COUNSELING & PSYCHOLOGICAL
SVCS CENTER
APPALACHIAN STATE UNIVERSITY
BOONE, NC 28608
704-262-3180

8963
MARGARET SKIDMORE
REGISTRAR
BELMONT ABBEY COLLEGE
100 BELMONT - MT. HOLLY RD
BELMONT, NC 28012
704-825-6732

9690
JEFF KORNEGAY
REGISTRAR/COUNSELOR
BLADEN COMMUNITY COLLEGE
PO BOX 266
DUBLIN, NC 28332
919-862-2164

8174
PAUL O'KEEFE
ACADEMIC COORDINATOR
CAMPBELL UNIVERSITY
PO BOX 8666
CAMP LEJEUNE, NC 08547
910-451-3097

9081
TOM HARRIS
DIRECTOR
CAMPBELL UNIVERSITY-FORT BRAGG
PO BOX 70659
FORT BRAGG, NC 28307
910-436-3242

8437
EDNA DAVIS JOHNSON
REGISTRAR
CARTERET COMMUNITY COLLEGE
3505 ARENDELL ST
MOREHEAD CITY, NC 28557-2989
919-247-4142

*9007
LINDA DIPPOLD
DEAN OF ACADEMIC AFFAIRS
CECILS COLLEGE
1567 PATTON AVE
ASHEVILLE, NC 28806
828-252-2486

***9692**
PHYLLIS HUFF
DIRECTOR OF HUMAN RESOURCES DEVELOPMENT
CENTRAL CAROLINA COMMUNITY COLLEGE
1105 KELLY DR
SANFORD, NC 27330
919-775-5401 EXT 324

***9693**
DOROTHY WALLACE
CHAIRPERSON, DEPARTMENT OF BUSINESS
CHOWAN COLLEGE
PO BOX 1848
MURFREESBORO, NC 27855
919-398-4101

9102
JOHN JAY
STUDENT AFFAIRS
COASTAL CAROLINA COMMUNITY COLLEGE
444 WESTERN BLVD
JACKSONVILLE, NC 28540
919-455-1221

9694
JUDITH LEE COTTRELL
DIRECTOR, ADMISSIONS
DAVIDSON COUNTY COMMUNITY COLLEGE
PO BOX 1287
LEXINGTON, NC 27293
336-249-8186

9131
MARION P. SYKES
DIRECTOR OF ADMISSIONS
EAST CAROLINA UNIVERSITY
FIFTH ST
GREENVILLE, NC 27858
919-757-6811

8379
BRUCE WOLFE
CENTER DIRECTOR
EMBRY-RIDDLE AERO UNIVERSITY
4th MSSQ/MSE
SJAFB, NC 27531
919-734-9211

***9145**
WESTRA J. MCGAHEE
TESTING SPECIALIST
FAYETTEVILLE STATE UNIVERSITY
134 ADMINISTRATION BLDG
1200 MURCHISON RD
FAYETTEVILLE, NC 28301
910-486-1302

9146
TOM GRAVES
DIRECTOR
FAYETTEVILLE TECHNICAL COMMUNITY COLLEGE
BOX 35236
FAYETTEVILLE, NC 28303
910-497-1112

***8581**
JOSEPH L. TILLMAN
TESTING COORDINATOR
JAMES SPRUNT COMMUNITY COLLEGE
PO BOX 398
KENANSVILLE, NC 28349
910-296-2506

***8880**
ROBERT L. BINGHAM
REGISTRAR
JOHN WESLEY COLLEGE
2314 NORTH CENTENNIAL ST
HIGH POINT, NC 27265-3197
336-889-2262

8423
JAMES STONESIFER
ACADEMIC AFFAIRS
LEES-McRAE COLLEGE
PO BOX 128
BANNER ELK, NC 28604
704-898-5241

***9206**
C. EDWIN BROWN, JR.
ACADEMIC DEAN
LOUISBURG COLLEGE
501 NORTH MAIN ST
LOUISBURG, NC 27549
919-496-2521

***9222**
CORINTHIA ATTAWAY
ADMINISTRATIVE ASSISTANT
METHODIST COLLEGE
5400 RAMSEY ST
FAYETTEVILLE, NC 28311
910-630-7251

***9015**
DEBORAH WRIGHT
DIRECTOR OF CHARLOTTE CAMPUS
MONTREAT COLLEGE - CHARLOTTE CAMPUS
4135 SOUTH STREAM BLVD, STE 200
CHARLOTTE, NC 28217
704-357-3390

***9121**
MARK GIBBS
DIRECTOR OF STUDENT SERVICES
MONTREAT COLLEGE
SCHOOL OF PROFESSIONAL AND ADULT STUDIES
318 RIDGEFILED CT
MONTREAT, NC 28806
704-667-5044

***8372**
MR. DAVID L. BOURGEOIS
ASSISTANT REGISTRAR
MOUNT OLIVE COLLEGE
634 HENDERSON ST
MOUNT OLIVE, NC 28365
919-658-7165

***8288**
CRYSTAL Y. GHOLSTON
PROGRAM REPRESNTATIVE
MOUNT OLIVE COLLEGE
2131 S GLENBURNIE RD, SUITE 6
NEW BERN, NC 28562
919-633-8470

***8241**
CHRISTIE STRICKLAND
TEST ADMINISTRATOR
MOUNT OLIVE COLLEGE
PO BOX 12142
RESEARCH TRIANGLE PARK, NC 27709-2142
919-941-2970

***8294**
MARY DUKE BARNWELL
PROGRAM REPRESENTATIVE
MOUNT OLIVE COLLEGE-WILMINGTON
1422 COMMONWEALTH DR
WILMINGTON, NC 28403
910-256-0255

***9434**
WILLIAM S. CARVER
DEAN, CONTINUING EDUCATION
NASH COMMUNITY COLLEGE
OLD CARRIAGE ROAD
PO BOX 7488
ROCKY MOUNT, NC 27804-0488
919-443-4011

***9251**
CAROL NICHOLS
TRANSFER COORDINATOR
NORTH CAROLINA STATE UNIVERSITY
112 CEELE HALL, PO BOX 7103
RALEIGH, NC 27695-7103
919-515-2434

***9252**
SHARRON McNEIL-DAVIS
SITE COORDINATOR FOR DANTES TESTING
NORTH CAROLINA WESLEYAN COLLEGE
3400 NORTH WESLEYAN BLVD
ROCKY MOUNT, NC 27804
252-985-5128

***8163**
JENNIFER MUMMEY
TEST ADMINISTRATOR
NORTH CAROLINA WESLEYAN COLLEGE - RALEIGH BRANCH
4900 WATERS EDGE DR, SUITE 260
RALEIGH, NC 27606
919-851-2130

9701
LAWRENCE DURRETT
REGISTRAR
PFEIFFER UNIVERSITY
MISENHEIMER, NC 28109
704-463-1360, X2057

***8473**
JACKSON ELLIOTT
VP, ADULT COMMUNITY AFFAIRS
PIEDMONT COMMUNITY COLLEGE
PO BOX 1197
ROXBORO, NC 27573
919-599-1181

***8377**
PATSYANNA BARKER
CHIEF GED EXAMINER
RANDOLPH COMMUNITY COLLEGE
PO BOX 1009
ASHBORO, NC 27204-1009
910-633-0241

This directory lists colleges & universities that award credit for passing scores on DSSTs.

*** INSTITUTIONS THAT ADMINISTER DSSTs**

**** Some schools may restrict testing to students admitted to or enrolled in the institution or offer a limited testing schedule.**

Students should contact the test center to determine current policies and schedules for DANTES testing.

*8132
REG BOLAND
DIR, COUNSELING & ASSESSMENT
ROWAN-CABARRUS COMMUNITY COLLEGE
PO BOX 1595
SALISBURY, NC 28145-1595
704-637-0760

9708
JACKIE CLARK
DIRECTOR OF ADMISSIONS
UNIV. OF NORTH CAROLINA AT PEMBROKE
PO BOX 1510
PEMBROKE, NC 28372
910-521-6262

*8286
MARTHA SANBURG
COORD, ACADEMIC ASSMT
UNIVERSITY OF NORTH CAROLINA-CHARLOTTE
9201 UNIVERSITY CITY BLVD
330-F FRETWELL BUILDING
CHARLOTTE, NC 28223
704-547-2213

*8322
KAREN DORAN
TESTING COORDINATOR
CNSLING & PSYCHOLOGICAL CTR
UNIVERSITY OF NC-CHAPEL HILL
CB #7471
UNC-CH
CHAPEL HILL, NC 27599-7471
919-962-2316

8969
JIM REED
DIRECTOR OF ADMISSIONS
WESTERN PIEDMONT COMMUNITY COLLEGE
1001 BURKEMONT AVE
MORGANTON, NC 28655
704-438-6000

NORTH DAKOTA

*8804
MARSHALL MELBYE
REGISTRAR
DICKINSON STATE UNIVERSITY
CAMPUS DR
DICKINSON, ND 58601
701-227-2331

9192
ARLO D STEVICK
DEAN OF VOCATIONAL TECHNICAL EDUCATION
LAKE REGION JUNIOR COLLEGE
NORTH COLLEGE DR
DEVILS LAKE, ND 58301
701-662-8683

8569
RICHARD HOLM
ADMISSIONS/RECORDS
NORTH DAKOTA STATE COLLEGE
800 NORTH 6TH ST
WAHPETON, ND 58076
701-671-1130

8658
RICHARD PETERSON
ACAD. OUTREACH OFFICER
UNIVERSITY OF NORTH DAKOTA-LAKE REGION
319 MSS/DPE BLDG 252
344 6th AVE
GRAND FORKS AFB, ND 58205-6336
701-594-8192

*9213
KRIS MAGSTADT
DIRECTOR
CAREER SVCS & TESTING
UNIVERSITY OF MARY
7500 UNIVERSITY DR
BISMARCK, ND 58504-9652
1-800-408-6279 EXT 414

OHIO

*8799
SALLY A HABIB
REGISTRAR
THE McGREGOR SCHOOL
ANTIOCH UNIVERSITY
800 LIVERMORE
YELLOW SPRINGS, OH 45387-1609
513-767-6321

9716
TINA MARTINI
DIR, TRANSFER EVALUATION
BOWLING GREEN STATE UNIVERSITY
110 ADMINISTRATION BLDG
BOWLING GREEN, OH 43403
419-372-7963

9888
YOLANDA LAWRENCE
COORD, VETERANS AFFAIRS
CINCINNATI TECHNICAL & COMMUNITY COLLEGE
3520 CENTRAL PKWY
CINCINNATI, OH 45223
513-569-1543

*8868
PATTI REID
PROGRAM COORDINATOR
MANAGEMENT NURSING SERVICES
COLLEGE OF MOUNT ST. JOSEPH
127 SOUTH DAVIS AVE
COLUMBUS, OH 43222
614-225-5261

*8640
SUNITHA NARAYEN
COORDINATOR OF EXPERIENTIAL LEARNING
COLLEGE OF MOUNT ST. JOSEPH
5701 DELHI RD
CINCINNATI, OH 45233
513-244-4712

8755
JAMES HAMBERG
DIRECTOR OF RECORDS AND REGISTRATION
COLUMBUS STATE COMMUNITY COLLEGE - COLUMBUS
550 EAST SPRING ST
COLUMBUS, OH 43215
614-227-2635

*9338
RENA WHEELER
DUBLIN SITE COORDINATOR
COLUMBUS STATE COMMUNITY COLLEGE - DUBLIN
6190 SHAMROCK COURT, SUITE 100
DUBLIN, OH 43016
614-761-2800

*9719
CAROL L. THOMAS
DIRECTOR
DEFIANCE COLLEGE
701 NORTH CLINTON
DEFIANCE, OH 43512
419-783-2366

9020
SUZANNE LYBARGER
COORDINATOR, ADVISING AND ASSESSMENT
EDISON COMMUNITY COLLEGE
1973 EDISON DR
PIQUA, OH 45356
513-778-8600

9027
CHARLES MASCELLINO
DIRECTOR OF ADMISSIONS
JEFFERSON TECHNICAL COLLEGE
4000 SUNSET BLVD
STEUBENVILLE, OH 43952
614-264-5591

*8417
DEBBIE TORRES
CAREER SERVICES CENTER
KENT STATE UNIVERSITY
261 MICHAEL SCHWARTZ CENTER
KENT, OH 44242
216-672-2867

*8490
ARTHUR J. ACTON
TESTING COORDINATOR
MARIETTA COLLEGE
MARIETTA, OH 45750
614-373-4643

*8035
BRENDA HYSON
DIRECTOR OF TESTING
MOUNT VERNON NAZARENE COLLEGE
800 MARTINSBURG RD
MOUNT VERNON, OH 43050
740-397-6862 EXT 4230

*8359
CAROLYN LEARNED
ADMISSIONS & REGISTRATION COORDINATOR
MT. VERNON NAZARENE COLLEGE
EXECUTIVE CENTER FOR LIFELONG LEARNING
2000 POLARIS PKWY SUITE 130
COLUMBUS, OH 43240-2007
740-397-6862 EXT 6003 OR 6004

8641
DAVID rand
COORDINATOR OF LEARNING
NORTHWEST TECHNICAL COLLEGE
BOX 246-A ROUTE 1
ARCHBOLD, OH 43502
419-267-5511

9263
SHIRLEY NUCKLOS
REGISTRAR
OHIO DOMINICAN COLLEGE
1216 SUNBURY RD
COLUMBUS, OH 43219
614-253-2741

***9264**
NANCY L. DOWNEY
TEST ADMINISTRATOR
TESTING DEPARTMENT
OHIO STATE UNIVERSITY
820 LINCOLN TOWER
1800 CANNON DR
COLUMBUS, OH 43210
614-292-2241

8402
SUSAN WESTON
ADMINISTRATOR
PARK COLLEGE
88 MSS/DPEEE
2000 ALLBROOK DR
SUITE 3, RM 113
WRIGHT-PATTERSON AFB, OH 45433-5315
513-257-2936

9309
CAROLYN MANN
COOR, ACADEMIC ASSMT CTR
SINCLAIR COMMUNITY COLLEGE
444 W THIRD ST
DAYTON, OH 45402
937-512-2940

8951
SHARON PURVIS
REGISTRAR
SOUTHERN STATE COMMUNITY COLLEGE
200 HOBART DR
HILLSBORO, OH 45133
513-393-3431

8900
JEAN J. DAIL
VETERANS COORDINATOR
UNIVERSITY OF TOLEDO
2801 WEST BANCROFT
TOLEDO, OH 43606
419-537-2701

8978
PATRICIA A. BOWER
ADULT DEGREE PROGRAMS
TIFFIN UNIVERSITY
155 MIAMI ST
TIFFIN, OH 44883
419-447-6442

***9347**
ROBIN V. PELTON
COORDINATOR/TESTING
UNIVERSITY OF AKRON
257 CARROLL ST
AKRON, OH 44325-4303
330-972-7084

***8259**
JERRY LESSARD
DIR OF ADVISING-IDEAL PROGRAM
WALSH UNIVERSITY
2623 SOUTH ARLINGTON RD
AKRON, OH 44319-2044
330-490-7293

9395
GAIL FRED
REGISTRAR
WRIGHT STATE UNIVERSITY
3640 COLONEL GLENN HWY
DAYTON, OH 45435-0001
937-775-5588

OKLAHOMA

***8411**
RUTH ANN COLAW
TESTING DIRECTOR
BARTLESVILLE WESLEYAN COLLEGE
2201 SILVER LAKE RD
BARTLESVILLE, OK 74006
918-333-6151

***9080**
MELISSA DEES-DELGADO
DIRECTOR OF TESTING
CAMERON UNIVERSITY
2800 WEST GORE
LAWTON, OK 73505
405-581-2212

8625
JANET E. BAHR
ACADEMIC AFFAIRS
NORTHEASTERN STATE UNIVERSITY
TAHLEQUAH, OK 74464
918-456-5511

***8583**
SUE FISHER
TESTING COORDINATOR
NORTHERN OKLAHOMA COLLEGE
1220 EAST GRAND AVE
TONKAWA, OK 74653
405-628-6357

9660
JUDY KINNAMON
TESTING COORDINATOR
OKALAHOMA BAPTIST UNIVERSITY
500 W. UNIVERSITY
SHAWNEE, OK 74801-2590
405-275-2850

***8526**
STEVE BAILEY
DIRECTOR
PACE PROGRAM
OKLAHOMA CHRISTIAN UNIVERSITY
OKLAHOMA CITY, OK 73136
1-877-345-PACE or 405-425-1900

8822
GLORIA CARDENAS BARTON
REGISTRAR
OKLAHOMA CITY COMMUNITY COLLEGE
7777 SOUTH MAY AVE
OKLAHOMA CITY, OK 73159
405-682-1611

8347
DENISE SHORT
DIRECTOR
OKLAHOMA CITY UNIVERSITY
PRIOR LEARNING AND UNIV. STUDIES
2501 NORTH BLACKWELDER
OKLAHOMA CITY, OK 73106
405-521-5288

9968
BARBARA COX
DIR. ACADEMIC ASSESSMENT
REDLANDS COMMUNITY COLLEGE
1300 COUNTRY CLUB RD
BOX 370
EL RENO, OK 73036
405-262-2552, X2609

9267
EVELYN HUTCHINGS
REGISTRAR/DIRECTOR OF ADMISSIONS & RECORDS
ROSE STATE COLLEGE
6420 SOUTHEAST 15th ST.
MIDWEST CITY, OK 73110-2799
405-736-0203

8652
SYDNEY FARR
DIRECTOR OF ADULT EDUCATION
TRI COUNTY TECH
6101 NOWATA RD
BARTLEVILLE, OK 74006

8828
BARBARA CHILDERS
REGISTRAR
TULSA JUNIOR COLLEGE
3727 EAST APACHE
TULSA, OK 74115
918-834-5071

***9087**
LYNDA D. POE
DIRECTOR, TESTING CENTER
UNIVERSITY OF CENTRAL OKLAHOMA
100 NORTH UNIVERSITY DR, RM 185
EDMOND, OK 73034-5209
405-974-2387

9366
JACK HUDSON
DIRECTOR OF ADMISSIONS
UNIVERSITY OF SCIENCE & ARTS OF OKLAHOMA
17th & GRAND, PO BOX 82345
CHICKASHA, OK 73018
405-224-3140

***9362**
SUSAN MACMILLAN
TEST ADMINISTRATOR
UNIVERSITY OF OKLAHOMA
1600 SOUTH JENKINS, RM 101
NORMAN, OK 73072-6400
405-325-1208

This directory lists colleges & universities that award credit for passing scores on DSSTs.

*** INSTITUTIONS THAT ADMINISTER DSSTs**
**** Some schools may restrict testing to students admitted to or enrolled in the institution or offer a limited testing schedule.**

Students should contact the test center to determine current policies and schedules for DANTES testing.

OREGON

9740
MARY BEZODIS
EVALUATIONS SPECIALIST
CLACKAMAS COMMUNITY COLLEGE
19600 SOUTH MOLALLA AVE
OREGON CITY, OR 97045-7998
503-657-6958

8653
DEA L. WELLS
REGISTRAR
EASTERN OREGON STATE COLLEGE
1410 L AVE
LA GRANDE, OR 97850
503-962-3392

9193
JOHN WINQUIST
COUNSELING
LANE COMMUNITY COLLEGE
4000 EAST 30th AVE
EUGENE, OR 97405
503-726-2204

8808
KENNETH WILLIAMS
REGISTRAR
LINFIELD COLLEGE
McMINNVILLE, OR 97128
503-472-4121

*9743
LONNA LIDDELL
PROGRAM ASSISTANT
MARYLHURST UNIVERSITY
LEARNING ASSESSMENT CENTER
17600 PACIFIC HIGHWAY
PO BOX 261
MARYLHURST, OR 97036-0261
800-634-9982 EXT. 3326

*8987
DR. ALLEN BELCHER
DEGREE COMPLETION PROGRAM
NORTHWEST CHRISTIAN COLLEGE
828 EAST ELEVENTH AVE
EUGENE, OR 97401
503-343-1641

9438
LINDA S. NEWELL
REGISTRAR
OREGON INSTITUTE OF TECHNOLOGY
3201 CAMPUS DRIVE
KLAMATH FALLS, OR 97601
541-885-1143

8010
JOHN SNODGRASS
OFFICE OF REGISTRAR
PACIFIC UNIVERSITY
2043 COLLEGE WAY
FOREST GROVE, OR 97116
503-357-6151

*9747
TED RISSER
DIRECTOR, ENROLLMENT SERVICES
ROGUE COMMUNITY COLLEGE
3345 REDWOOD HWY
GRANTS PASS, OR 97527
541-471-3500

9976
LARRY SHIPLEY
DIRECTOR OF ADMISSIONS
UMPQUA COMMUNITY COLLEGE
PO BOX 967
ROSEBURG, OR 97470
503-440-4604

*9961
DAVID ESPINOZA
TESTING OFFICE
UNIVERSITY OF OREGON
1590 EAST 13th AVE, RM 238
EUGENE, OR 97403
541-346-3230

*8305
BECKY LEITHOLD
ASSESSMENT COUNSELOR
WESTERN BAPTIST COLLEGE
5000 DEER PARK DR, SE
SALEM, OR 97301
503-375-7590

PENNSYLVANIA

9749
BEKI STEIN
REGISTRAR
ALVERNIA COLLEGE
READING, PA 19607
610-796-8200

8747
ELIZABETH VERMEY
DIRECTOR OF ADMISSIONS
BRYN MAWR COLLEGE
BRYN MAWR, PA 19010
215-526-5000

*9925
JEAN RYERSBACH
COORDINATOR
BUCKS COUNTY COMM COLLEGE
BLDG SU-12, SWAMP RD
NEWTOWN, PA 18940
215-968-8476

*8319
LORI KAHN
COORDINATOR ASSESSMENT
PRIOR LEARNING
CABRINI COLLEGE
610 KING OF PRUSSIA RD
RADNOR, PA 19087
610-902-8296

*8758
JEANNINE METAL
ASST DIRECTOR OF CAREER SVCS
CALIFORNIA UNIVERSITY OF PENNSYLVANIA
250 UNIVERSITY AVE
CALIFORNIA, PA 15419
412-938-4413

9889
BARBARA HEGEL
REGISTRAR
CEDAR CREST COLLEGE
100 COLLEGE DR
ALLENTOWN, PA 18104
610-437-4771

9115
SHERRY ROSE-BOND
DIRECTOR OF TESTING CENTER
COMM COLL OF PHILADELPHIA
1700 SPRING GARDEN ST
PHILADELPHIA, PA 19130
215-751-8000

9122
LINDA LONG
DIRECTOR, ASSESSMENT CENTER
DELAWARE COUNTY COMMUNITY COLLEGE
RT 252 AND MEDIA LINE
MEDIA, PA 19063
215-359-5000

*8320
ELEANOR PRICE
MATH & PHYSICS DEPT.
DELAWARE VALLEY COLLEGE
700 EAST BUTLER AVE
DOYLESTOWN, PA 18901
215-345-1500

8834
DIANE S.H. BACCI
REGISTRAR
EASTERN COLLEGE
1300 EAGLE ROA
ST. DAVIDS, PA 19087-3696
610-341-5854

*9528
JOHN DAVID BOWMAN
DIRECTOR OF LANCASTER PROGRAMS
EASTERN MENNONITE UNIVERSITY
1770 HEMPSTEAD ROAD
PO BOX 10936
LANCASTER, PA 17605-0936
717-397-5190

8123
JANE LIOTTA
REGISTRAR
FRANKLIN & MARSHALL COLLEGE
PO BOX 3003
LANCASTER, PA 17604-3003
717-291-3975

*8836
JOHN T. PROSPERI
DIRECTOR, CONTINUING ED
GREATER ALTOONA CAREER & TECH CENTER
1500 4th AVE
ALTOONA, PA 16602
814-946-8469

8768
JOSEPH AMPREY, JR.
DEAN OF ACADEMIC SVCS
KUTZTOWN UNIVERSITY
KUTZTOWN, PA 19530
610-683-4215

9196
LESLIE LEVINE
ACADEMIC ADVISING
LA ROCHE COLLEGE
9000 BABCOCK BLVD
PITTSBURGH, PA 15237
412-367-9300

9755
DIANE LANG
REGISTRAR
LACKAWANNA JUNIOR COLLEGE
901 PROSPECT AVE
SCRANTON, PA 18505
717-961-7810

8983
DR. PETER W. TEAGUE
DEAN OF UNDERGRADUATE STUDIES
LANCASTER BIBLE COLLEGE
901 EDEN RD
LANCASTER, PA 17601
717-569-7071

9197
BARBARA J. DENISON
ACADEMIC ADVISER
LEBANON VALLEY COLLEGE
101 NORTH COLLEGE AVE
ANNVILLE, PA 17003
717-867-6213

*9757
MARGARITA RODRIGUEZ
OFFICE ASSISTANT/SECRETARY
LEHIGH CARBON COMMUNITY COLLEGE
4525 EDUCATION PARK DR
SCHNECKSVILLE, PA 18078
610-799-1137

*9209
BARBARA PRICE
DIRECTOR OF CAREER PLANNING
LUZERNE COUNTY COMMUNITY COLLEGE
1333 SOUTH PROSPECT ST
NANTICOKE, PA 18634
717-829-7456

9216
ANN BOLAND-CHASE
REGISTRAR
MARYWOOD UNIVERSITY
2300 ADAMS AVE
SCRANTON, PA 18509-1598
570-348-6280

*8817
LINDA KVAMME
ASST DEAN, CONTG STUDIES
MORAVIAN COLLEGE
1200 MAIN ST
BETHLEHEM, PA 18018
610-861-1384

8312
VICTORIA FISHER
DIRECTOR, CONTINUING ED
NEUMANN COLLEGE
ASTON, PA 19014
215-558-5530

*9471
LARRY MELLEY
DEAN
CONTINUING EDUCATION
NEWPORT BUSINESS INSTITUTE
941 WEST THIRD STREET
WILLIAMSPORT, PA 17701
717-322-0369

*9971
WALTON B. HILL
TEST ADMINISTRATOR
PENNSYLVANIA INSTITUTE OF TECHNOLOGY
800 MANCHESTER AVE
MEDIA, PA 19063
610-892-1505

8953
EDWIN ESCALET
DIRECTOR OF ADMISSIONS
PENNSYLANIA STATE UNIVERSITY-HARRISBURG
777 W HARRISBURG PK
MIDDLETOWN, PA 17057
717-948-6100

9277
SCOTT HEALY
DEAN OF ADMISSIONS
PENNSYLANIA STATE UNIVERSITY
2010 OLD MAIN
UNIVERSITY PK, PA 16802
814-865-4700

*8266
NANCY KARLIK
DIRECTOR
PENNSYLVANIA STATE UNIVERSITY HARRISBURG CENTER
234 NORTH THIRD ST
HARRISBURG, PA 17101
717-783-0433

*8022
JIM TOBIN
DIRECTOR, PART-TIME PROGRAMS
PIERCE COLLEGE
1420 PINE ST
PHILADELPHIA, PA 19102
215-545-6400

*9736
SUSAN SHEA
COORD, FACULTY & STAFF EDUC.
PITTSBURGH TECHNICAL INST.
635 SMITHFIELD ST
PITTSBURGH, PA 15222
412-471-1011

8468
FAYE ROSS
DIRECTOR, LEARNING CENTER
PHILADELPHIA COLLEGE OF TEXTILES AND SCIENCE
SCHOOL HOUSE LANE & HENRY AVE
PHILADELPHIA, PA 19144
215-951-2734

*9760
JANET AGNEW
TEST ADMINISTRATOR
ACADEMIC SERVICES
LOWER LEVEL, TH
POINT PARK COLLEGE
201 WOOD ST
PITTSBURGH, PA 15222-1984
412-392-3864

9284
SCOTT RICHARDSON
DIRECTOR, ASSESSMENT CTR.
READING AREA COMMUNITY COLLEGE
10 SOUTH SECOND ST
READING, PA 19603
215-372-4721

9983
JANICE BABIN
DIRECTOR OF ADMISSIONS
RETS EDUCATION CENTER
WESTCHESTER PK & MAIN RD
BROOMALL, PA 19008
215-353-7630

9761
FRANCIS PERRY
DIRECTOR, SUPPORT SERVICES
ACADEMIC AFFAIRS
ROBERT MORRIS COLLEGE
NARROWS RUN RD
CORAPOLIS, PA 15108
412-262-8200

*9013
MARY STANTON
OFFICE MANAGER
ROSEMONT COLLEGE
1400 MONTGOMERY AVE
ROSEMONT, PA 19010-1699
610-526-2955

*9763
MARTHA RAAK
DEAN, CONTINUING EDUCATION
SETON HILL COLLEGE
SETON HILL DR
GREENSBURG, PA 15601
724-838-4208

8946
RYTA MIRISCIOTTI
ADMIN COORD ACADEMIC RECORDS
SLIPPERY ROCK UNIVERSITY
107 OLD MAIN
SLIPPERY ROCK, PA 16057
724-738-2010

9292
FATHER TOM BOURQUE
CONTINUING EDUCATION
ST. FRANCIS COLLEGE
LORETTO, PA 15940
814-472-3012

8310
RICHARD O'MALLEY
DIRECTOR
ST. JOSEPH'S UNIVERSITY
5600 CITY AVE
PHILADELPHIA, PA 19131
215-660-1000

8341
MIKE SCHRETTENBRUNNER
SENIOR ACADEMIC ADVISOR
STRAYER COLLEGE - LOUDOUN CAMPUS
45150 RUSSELL BRANCH PKWY
ASHBURN, PA 22011
703-729-8815

8982
SHIRLEY M. ADAMS
DEAN, DEXTER HANLEY COLLEGE
UNIVERSITY OF SCRANTON
SCRANTON, PA 18510
717-942-7580

8977
DONALD S. ROWE
DEAN
VALLEY FORGE MILITARY COLLEGE
1001 EAGLE RD
WAYNE, PA 19087
215-688-1800

8873
LINDA KAMINSKI
ACADEMIC AFFAIRS
WESTMORELAND COUNTY COMMUNITY COLLEGE
YOUNGWOOD, PA 15697
412-925-4212

8651
GUNLOG ANDERSON
DIRECTOR, ADULT LEARNING PROGRAM
WILSON COLLEGE
CHAMBERSBURG, PA 17201
717-264-4141

9381
ROBERT DANNER
ADMISSIONS
YORK COLLEGE OF PENNSYLANIA
COUNTRY CLUB RD
YORK, PA 17405-7199
717-849-1600

RHODE ISLAND

*9767
RONALD SCHERTZ
DEAN OF STUDENTS
COMM COLL OF RHODE ISLAND
400 EAST AVE
WARWICK, RI 02886
401-825-1000

9183
JOANNE M. GALENSKI
REGISTRAR
JOHNSON & WALES UNIVERSITY
8 ABBOTT PARK PLACE
PROVIDENCE, RI 02903
401-456-1088

*9286
AGNES J. ALBRIGHT
CENTER FOR EVALUATIONS& RESEARCH
RHODE ISLAND COLLEGE
600 MT. PLEASANT AVE
PROVIDENCE, RI 02908
401-456-8000

8401
JOHN STOUT
DEAN, SCHOOL OF CONTINUING ED
ROGER WILLIAMS UNIVERSITY
ONE OLD FERRY RD
BRISTOL, RI 02809
401-253-1040

SOUTH CAROLINA

9052
TERRIE MARSHAL
DIRECTOR OF ADMISSIONS
BAPTIST COLLEGE AT CHARLESTON
JONES HALL-BOX 118087
CHARLESTON, SC 29423-8087
803-863-8060

*9061
PARYN MOSS
COORDINATOR OF TESTING
BENEDICT COLLEGE
HARDEN & BLANDING STS
COLUMBIA, SC 29204
803-253-5207

9335
DEBORAH C. CALHOUN
REGISTRAR, DIR OF STUDENT INFORMATION
CENTRAL CAROLINA TECHNICAL COLLEGE
506 NORTH GUIGNARD DR
SUMTER, SC 29150
803-778-7857

8955
MARY K. NEWTON
ASSISTANT DEAN OF STUDENTS
CHESTERFIELD-MARLBORO TECHNICAL COLLEGE
PO DRAWER 1007
CHERAW, SC 29520
843-921-6935

8441
GEORGE HOGAN
DIRECTOR
COKER COLLEGE
ARMY EDUCATION CENTER
FT. JACKSON, SC 29207
803-738-9299

9006
MARY JO BLACK
REGISTRAR
HORRY-GEORGETOWN TECHNICAL COLLEGE
HWY 501 EAST, PO BOX 261966
CONWAY, SC 29528-6066
843-349-5246

*9773
C.R. HORTON
DIRECTOR FOR THE BLOCK PROGRAM
LIMESTONE COLLEGE
1115 COLLEGE DR
GAFFNEY, SC 29340
800-795-7151 EXT 586

*8315
EMMA PEASANT
TEST COORDINATOR
STUDENT ASSESSMENT - AIRPORT CAMPUS
MIDLANDS TECHNICAL COLLEGE
1260 LEXINGTON DR
WEST COLUMBIA, SC 29170
803-822-3661

*8014
LIZ BELL, CHAIR
DIVISION OF GENERAL STUDIES
MORRIS COLLEGE
100 WEST COLLEGE ST
SUMTER, SC 21950
803-775-9371

*9002
WILLIAM GASH
REGISTRAR
PRESBYTERIAN COLLEGE
CLINTON, SC 29325
803-833-2820

9427
GRETCHEN MUNROE
DIRECTOR, SHAW
SAINT LEO COLLEGE
522 NELSON, BLDG #800
SHAW AFB, SC 29152
803-666-3221

8135
JUDY BLACK
ASSOCIATE REGISTRAR
SOUTHERN WESLEYAN UNIVERSITY
ONE WESLEYAN DR
PO BOX 1020, SWU BOX 495
CENTRAL, SC 29630
864-639-2453

9056
FRANCES BOGAN
MILITARY COORDINATOR
TECHNICAL COLLEGE OF THE LOW COUNTRY
921 SOUTH RIBAUT RD
PO BOX 1288
BEAUFORT, SC 29901
803-525-8230

9342
LaVERNE B. BAILEY
STUDENT AFFAIRS
TRIDENT TECHNICAL COLLEGE
PO BOX 118067
CHARLESTON, SC 29411
803-569-6549

8966
GORDON A. CREWS
DIR, CRIMINAL JUSTICE AND MILITARY PROGRAMS
UNIVERSITY OF SOUTH CAROLINA
801 CARTERET ST
BEAUFORT, SC 29902
803-522-7497

9367
SALLY BOYD
DEAN FOR LIFELONG LEARNING
UNIVERSITY OF SOUTH CAROLINA
900 ASSEMBLY ST
COLUMBIA, SC 29208
803-777-8155

9368
MARSH MYERS, JR.
REGISTRAR
USC-COASTAL CAROLINA COLLEGE
PO BOX 1954
CONWAY, SC 29526
803-448-1481

SOUTH DAKOTA

8435
CAROL THOMAS
BLACK HILLS STATE UNIVERSITY
ELLSWORTH BRANCH CAMPUS
ELLSWORTH AFB, SD 57706
605-923-5893

9065
APRIL M. MEEKER
ADMISSIONS AND RECORDS
BLACK HILLS STATE UNIVERSITY
1200 UNIVERSITY ST, USB 9502
SPEARFISH, SD 57799
605-642-6011

*9777
MARILYN HOLMGREN
ASSESSMENT DIRECTOR
NATIONAL AMERICAN UNIVERSITY
321 KANSAS CITY ST, PO BOX 1780
RAPID CITY, SD 57709
605-394-4839

8547
DR. RAY POSPISIL
REGISTRAR
UNIVERSITY OF SOUTH DAKOTA
VERMILLION, SD 57069
605-677-5011

TENNESSEE

*9050
BARBARA BLACKSTON
TESTING COORDINATOR
AUSTIN PEAY STATE UNIVERSITY
COLLEGE ST
CLARKSVILLE, TN 37044
615-648-6269

*8292
REBECCA WHEAT
COORDINATOR, STUDENT SVCS
BELHAVEN COLLEGE
5100 POPLAR AVE, SUITE 200
MEMPHIS, TN 38117
901-888-3345

*9779
SHIRLEY MARTIN
REGISTRAR
BETHEL COLLEGE
325 CHERRY ST
McKENZIE, TN 38201
901-352-4047

8798
L. EDWARD HART
REGISTRAR
CARSON-NEWMAN COLLEGE
CNC BOX 1985
JEFFERSON CITY, TN 37760
615-475-9061

*8162
BRIDGET REESE
SCHOOL OF ADULT EDUCATION
CRICHTON COLLEGE
3665 KIRBY PARKWAY
MEMPHIS, TN 38115
901-367-3892

*8579
DOUG HODGE
MANAGER OF ASSESSMENT
DYERSBURG STATE COMMUNITY COLLEGE
1510 LAKE RD
DYERSBURG, TN 38024
901-286-3359

*9067
JAMES H. MAXWELL
DIRECTOR OF TESTING
FREED HARDEMAN UNIVERSITY
158 EAST MAIN STREET
HENDERSON, TN 38340
901-989-6060

*8672
DR. VERNON EATON
VICE PRESIDEENT FOR ACADEMICS
JOHNSON BIBLE COLLEGE
7900 JOHNSON DR
KNOXVILLE, TN 37988
423-251-3436

9780
CONRAD DANIELS
DEAN OF ADMISSIONS & RECRUITMENT
LINCOLN MEMORIAL UNIVERSITY
CUMBERLAND GAP PKWY
HARROGATE, TN 37752
615-869-3611

*9781
JUDITH MALLORY
DIRECTOR
OFFICE OF COMMUNITY AND ADULT EDUCATION
MARYVILLE COLLEGE
502 E LAMAR ALEXANDER PKWY
MARYVILLE, TN 37804-5907
423-981-8052

*8705
TRACY BIOOTHE
ASST REGISTRAR/DIRECTOR OF TESTING
MILLIGAN COLLEGE
PO BOX 330
MILLIGAN, TN 37682
423-975-8037

9033
BILLY HIX
DIRECTOR OF ADMISSIONS
MOTLOW STATE COMMUNITY COLLEGE
PO 88100
TULLAHOMA, TN 37388
615-455-8511

8724
LEE BRUCKNER
ACADEMIC AFFAIRS
NASHVILLE BIBLE COLLEGE
PO BOX 78037
NASHVILLE, TN 37207
615-262-3125

*8477
SALLY ESSIN
NORTHEAST STATE TECHNICAL COMMUNITY COLLEGE
PO BOX 246
BLOUNTVILLE, TN 37617-0246

*8311
JOAN NEWMAN
DIRECTOR OF ACADEMIC ASSMT
PELLISSIPPI STATE TECHNICAL COLLEGE
10915 HARDING VALLEY RD
KNOXVILLE, TN 37933-0990
423-694-6543

8374
JOYCE K. PERRY
COORDINATOR OF ADMISSIONS
ROANE STATE COMMUNITY COLLEGE
ROUTE 8, BOX 69, PATTON LANE
HARRIMAN, TN 37748
615-354-3000

8679
SARAH C. BUFORD
REGISTRAR/DIR OF RECORDS
SHELBY STATE COMMUNITY COLLEGE
PO BOX 40568
MEMPHIS, TN 38174
901-528-6700

9878
K. R. DAVIS
DIRECTOR OF TESTING & COUNSELING
SOUTHERN MISSIONARY COLLEGE
DRAWER T
COLLEGEDALE, TN 37315

9322
JOHN CUMMINGS
DIRECTOR OF COUNSELING
STATE TECH INST AT MEMPHIS
5983 MACON COE
MEMPHIS, TN 38134
901-377-4111

*8723
RICHARD W. HOOPER
DIRECTOR OF TESTING
TENNESSEE STATE UNIVERSITY
330 10th NORTH, SUITE C
NASHVILLE, TN 37203-3401
615-963-7114

8954
LYDIA NORFLEET
REGISTRAR
TENNESSEE WESLEYAN COLLEGE
PO BOX 40
ATHENS, TN 37371
615-745-7504

*9784
PEGGY CARPENTER
DIRECTOR OF STUDENT ASSESSMENT
TREVECCA NAZARENE COLLEGE
333 MURFREESBORO RD
NASHVILLE, TN 37210
615-248-1268

*8863
SANDRA F. FORD
DIRECTOR OF STUDENT SERVICES
TUSCULUM COLLEGE
PO BOX 5689
GREENEVILLE, TN 37743
423-636-7330

This directory lists colleges & universities that award credit for passing scores on DSSTs.

*** INSTITUTIONS THAT ADMINISTER DSSTs**

**** Some schools may restrict testing to students admitted to or enrolled in the institution or offer a limited testing schedule.**

Students should contact the test center to determine current policies and schedules for DANTES testing.

*8593
CINDY SHIRLEY
DIRECTOR PRIOR LEARNING &
TESTING
UNION UNIVERSITY
1050 UNION UNIVERSITY DRIVE
JACKSON, TN 38305
901-661-5487

*9782
JAMES DUSTIN MCDANIEL
LICENSING EXAMINER
UNIVERSITY OF MEMPHIS
3770 DESOTO - ROOM 111
MEMPHIS, TN 38152
901-678-2428

TEXAS

8486
HELEN CARR
TESTING COORDINATOR
ALAMO COMMUNITY COLLEGE
1300 SAN PEDRO AVE
SAN ANTONIO, TX 78284
512-220-1500

8870
MARIE FERGUSON
GRADUATION ADVISOR
ALVIN COMMUNITY COLLEGE
3110 MUSTANG RD
ALVIN, TX 77511
713-388-4615

9041
REBECCA GALLOGLY
ACADEMIC COUNSELING
AMERICAN TECH UNIVERSITY
PO BOX 1416
KILLEEN, TX 76541

8245
CECILIA PADILLA-LARTEY
ADMISSIONS ASSISTANT II
AUSTIN COMMUNITY COLLEGE
5930 MIDDLE FISKVILLE RD
AUSTIN, TX 78752-4390
512-223-7762

9057
GERRY MORENO
TESTING COORDINATOR
BEE COUNTY COLLEGE
3800 CHARCO RD
BEEVILLE, TX 78102
512-358-3130

*8016
NIC SALAS
DIRECTOR OF TESTING
BILL J. PRIEST INSTITUTE
1402 CORINTH COURT ST,
SUITE 1670
DALLAS, TX 75215
214-860-5773/5774

*9088
GEORGE ERSKINE
TESTING DIRECTOR
CENTRAL TEXAS COLLEGE
U.S. HWY 190 WEST
KILLEEN, TX 76542
817-526-1254

9952
JOHN ATTANASIO
DIRECTOR, STUDENT SERVICES
PACIFIC FAR EAST CAMPUS
CENTRAL TEXAS COLLEGE (KOREA)
UNIT #15559
APO, AP 96283-0618
82-32-520-6807

*8614
WILLIAM R. CARVER
DIRECTOR , COUNSELING CENTER
COOKE COUNTY COLLEGE
PO BOX 815
GAINESVILLE, TX 76240
817-668-7731

8948
LARRY W. YEATER
UNIVERSITY REGISTRAR
CORPUS CHRISTI STATE UNIVERSITY
6300 OCEAN DR
CORPUS CHRISTI, TX 78412
512-994-6810

*8419
CHARLES HICKOX
DIRECTOR
DALLAS CHRISTIAN COLLEGE
2700 CHRISTIAN PKWY
DALLAS, TX 75234
214-484-6527

*8389
DEBORAH EDWARDS
COORDINATOR, TESTING
DEL MAR COLLEGE
101 BALDWIN BLVD
CORPUS CHRISTI, TX 78404
512-886-1200

*9008
CHARLES HELTON
DIRECTOR OF TESTING
EASTFIELD COLLEGE
3737 MOTLEY DR
MESQUITE, TX 75150
214-860-7010

*9142
CAROL GIARDANO
DIRECTOR, TESTING SVCS
EL PASO COMMUNITY COLLEGE
PO BOX 20500
EL PASO, TX 79998
915-594-2223

9791
DOROTHY KISER
REGISTRAR
HARDIN-SIMMONS UNIVERSITY
BOX 16190
ABILENE, TX 79698
915-670-1368

*9170
NANCY HASHEMI
TESTING DEPARTMENT
HOUSTON COMMUNITY COLLEGE
1300 HOLMAN ST, RM 201
HOUSTON, TX 77004
713-630-1115

8009
DON JACKSON
REGISTRAR
HOWARD PAYNE UNIVERSITY
1000 FISK HPU STATION BOX 836
BROWNWOOD, TX 76801
915-646-2502

*8307
JOHN CASTLEMAN
REGISTRAR
LUBBOCK CHRISTIAN UNIVERSITY
5601 19th ST
LUBBOCK, TX 79407-2099
806-293-742

9218
BOBBYE FRY
REGISTRAR
McMURRY UNIVERSITY
McMURRY STATION, BOX 338
ABILENE, TX 79697
915-691-6400

9874
CAMAL DAKIL
DEAN OF STUDENTS/ADMISSIONS
MIDLAND COLLEGE
3600 NORTH GARFIELD
MIDLAND, TX 79705
915-685-4500

9870
MARILYNN ADAMS
ASST to the REGISTRAR
MIDWESTERN STATE UNIVERSITY
3410 TAFT BLVD
WICHITA FALLS, TX 76308
817-689-4344

*8384
DEBBIE EBERLE
TESTING DEPT.
MOUNTAIN VIEW COLLEGE
4849 WEST ILLINOIS AVE
DALLAS, TX 75211
214-333-8778

*9243
WANDA GILLEN
COUNSELOR
NAVARRO COLLEGE
3200 WEST 7TH AVE
CORSICANA, TX 75110
214-874-6501

*8460
SUSAN LITTLEJOHN
DIRECTOR OF TESTING
NORTH LAKE COLLEGE
5001 NORTH MACARTHUR BLVD
IRVING, TX 75038
214-273-3161

*9793
DANIEL PEREZ
COORDINATOR OF ASSESSMENT
OUR LADY OF THE LAKE UNIVERSITY
411 S.W. 24th ST
SAN ANTONIO, TX 78207-4689
210-434-6711 EXT 337

8399
JAMES BOOFTER
ADMINISTRATOR
PARK COLLEGE
PO BOX 8039
FT. BLISS, TX 79908
915-562-8450

8298
AMY L. DEBRUICKER
RESIDENT CENTER ADMINITRATOR
PARK COLLEGE
47 MSS/DPE BLDG 316
540 SEOND ST., SUITE 1
LAUGHLIN AFB, TX 78843
830-298-3061

8829
DR. GEORGE STAFFORD
DIRECTOR OF TESTING
PRAIRIE VIEW UNIVERSITY
PO BOX 428
PRAIRIE VIEW, TX 77446
409-857-4611

8049
TERESA A. HARVEY
TESTING SUPERVISOR
SAM HOUSTON STATE UNIVERSITY
PO BOX 2418
HUNTSVILLE, TX 77341-2418
409-294-1221

8503
JIM WALKER
ASSOC DEAN OF TECHNICAL
& CONTINUING EDUCATION
SOUTH PLAINS COLLEGE
1401 SOUTH COLLEGE AVE
LEVELLAND, TX 79336
806-894-9611 EXT 2340

9320
STEPHEN B. SPRINGER
DIRECTOR, OCCUPATIONAL ED
SOUTHWEST TEXAS STATE UNIV.
SAN MARCOS, TX 78666
512-245-2115

8912
DR. JOHN A. COSTELLO
ASSOCIATE ACADEMIC VP
ST. EDWARD'S UNIVERSITY
3001 SOUTH CONGRESS AVE
AUSTIN, TX 78704
512-448-8581

9428
SARA HEYDON
DIRECTOR, INGLESIDE
SAINT LEO COLLEGE
125 CORAL SEA RD, #T9 BOX 622
INGLESIDE, TX 78362-0622
512-776-2808

*9330
JIM WILLIS
COORDINATOR, TESTING SVCS
ST. MARY'S UNIVERSITY
ONE CAMINO SANTA MARIA
SAN ANTONIO, TX 78228-8528
210-436-3135

*9331
PEARLENE CONYERS
COUNSELOR
ST. PHILLIP'S COLLEGE
2111 NEVADA ST
SAN ANTONIO, TX 78203
512-531-3508

*9334
DON ROBERTSON
CHIEF TEST ADMINISTRATOR
CAREER SVCS & TESTING OFFICE
SUL ROSS STATE UNIVERSITY
BOX C-171
ALPINE, TX 79832
915-837-8178

*8257
ANNA MACHALEK
TESTING COORDINATOR
TEMPLE COLLEGE
2600 SOUTH FIRST ST
TEMPLE, TX 76504
817-773-9961

*9795
HEIDI HALSTEAD
TESTING SVCS ADMINISTRATOR
TEXAS A & M UNIVERSITY
OO1 HISTORY BUILDING
COLLEGE STATION, TX 77843-4239
409-845-0532

*9790
MAXIMINO PLATA
DIRECTOR, STUDENT ASSESSMENT
TEXAS A & M UNIVERSITY
COMMERCE, TX 75429
903-886-5122

9797
STANLEY B. BLOCK
MJ NEELEY SCHOOL OF BUSINESS
TEXAS CHRISTIAN UNIVERSITY
TCU BOX 298530
FORT WORTH, TX 76129
817-921-7000

9337
KRISTIN PLAEHN
REGISTRAR
TEXAS LUTHERAN UNIVERSITY
1000 W COURT ST
SEGUIN, TX 78155
830-372-8040

*8491
SHERRY HOLLY
SUPERVISOR OF CONTINUING ED
TEXAS STATE TECHNICAL COLLEGE
307 NORTH BRECKENRIDGE AVE
BRECKENRIDGE, TX 76424
817-559-6556

*8474
SUSAN REEDER
TEST CENTER MANAGER
TYLER JUNIOR COLLEGE
1400 EAST DEVINE, PO BOX 9020
TYLER, TX 75711
903-510-2617

*9174
JULI GONZALEZ
COORDINATOR OF TESTING
UNIVERSITY OF THE INCARNATE WORD
4301 BROADWAY
SAN ANTONIO, TX 78209
210-829-3930

*9799
PATRICK DANIEL
DIR, LEARNING & ASSESSMENT
COUNSELING & TESTING SERVICE
UNIVERSITY OF HOUSTON
UNIVERSITY PARK
4800 CALHOUN ST
HOUSTON, TX 77204-3242
713-743-5454

8827
CHRISTINE WAGGONER
ADMISSIONS SUPERVISOR
UNIVERSITY OF NORTH TEXAS
PO BOX 13797 1401
W. PRAIRIE #305
DENTON, TX 76203
940-565-2681

*9558
KAREN KING
TESTING SUPERVISOR
UNIVERSITY OF TEXAS - AUSTIN
P.O. BOX 7700
AUSTIN, TX 78713-7700
512-471-7716

9805
DEBORAH GANT
TEST DIRECTOR
UNIVERSITY OF TEXAS AT DALLAS
PO BOX 830688
RICHARDSON, TX 75083
214-690-2111

9807
JOLEEN REYNOLDS
TESTING CENTER
UNIVERSITY OF TEXAS
6900 LOOP 1604 WEST
SAN ANTONIO, TX 78249
210-691-4125

8071
BRENDA S. KAYS
DIRECTOR OF COUNSELING
VERNON REGIONAL JUNIOR COLLEGE
4105 MAPLEWOOD AVE
WICHITA FALLS, TX 76308
817-696-8752

9384
VERNON L NORRIS
COORD, EXTERNAL PROGRAMS
WAYLAND BAPTIST UNIVERSITY
1900 WEST 7TH ST
PLAINVIEW, TX 79072
806-296-4708

8182
DILLARD WHITIS, PH.D
DIRECTOR, SAN ANTONIO CENTER
WAYLAND BAPTIST UNIVERSITY
1802 NE LOOP 410, SUITE 200
SAN ANTONIO, TX 78217
210-826-7595

This directory lists colleges & universities that award credit for passing scores on DSSTs.

*** INSTITUTIONS THAT ADMINISTER DSSTs**

**** Some schools may restrict testing to students admitted to or enrolled in the institution or offer a limited testing schedule.**

Students should contact the test center to determine current policies and schedules for DANTES testing.

UTAH

8812
JAN YOUNG, DIRECTOR
ADMISSIONS & RECORDS
COLLEGE OF EASTERN UTAH
451 EAST 400 NORTH
PRICE, UT 84501
801-637-2120

*8764
KRAIG MUNZERT
CURRICULUM COORDINATOR
HAWTHORNE UNIVERSITY
2965 EAST 3435 SOUTH
SALT LAKE CITY, UT 84109
801-485-1801

8805
GEOFFREY BRUGGER
DEAN, CONTINUING EDUCATION
SALT LAKE COMMUNITY COLLEGE
PO BOX 30808
SALT LAKE CITY, UT 84130
801-967-0365

*9924
DEBRA MARIN
TEST ADMINISTRATOR
UNIVERSITY OF UTAH
201 S. 1460 E - RM 498 SSB
SALT LAKE CITY, UT 84112
801-581-8733

8482
DENISE HODGKIN
DIRECTOR OF ASSESSMENT
ASSESSMENT CENTER - 120
UTAH VALLEY STATE COLLEGE
800 W 1200 SOUTH
OREM, UT 84058-5999
801-222-8269

9810
WAYNE WOOLSTON
REGISTRAR
WESTMINSTER COLLEGE OF SLC
1840 SOUTH 13th EAST
SALT LAKE CITY, UT 84105
801-484-7651

VERMONT

*8270
MARTHA ZIPF
DIRECTOR, ACCEL PROGRAM
COLLEGE OF ST. JOSEPH
71 CLEMENT RD
RUTLAND, VT 05701
802-773-5900

*8533
BETH ROBERTS
TRANSFER COORDINATOR
COMMUNITY COLLEGE OF VERMONT
PO BOX 120
WATERBURY, VT 05676
802-241-1190

8975
MARY G. VAN ARSDALE
DIRECTOR OF ADMISSIONS
SOUTHERN VERMONT COLLEGE
MONUMENT AVE, EXT
BENNINGTON, VT 05201
802-442-5427

*8517
KELLY JANE THOMAS
ACADEMIC ADVISOR
TRINITY COLLEGE
208 COLCHESTER AVE
BURLINGTON, VT 05401
802-658-0337

*8746
KAREN BEARDSLEY
MANAGER
VERMONT STATE COLLEGE
68 MAIN ST
SPRINGFIELD, VT 05156-2905
802-885-5183

VIRGINIA

8131
CHERYLL J. FRAZIER
DIRECTOR OF UNDERGRADUATE ADMISSIONS
AMERICAN MILITARY UNIVERSITY
9104-P MANASSAS DR
MANASSAS PARK, VA 22111
703-330-5398.

9051
KESHIA BAILEY
COORDINATOR OF STUDENT AFFAIRS
AVERETT COLLEGE
1593 SPRING HILL RD
VIENNA, VA 22182
703-448-0371

*8759
HELEN NAFZINGER
DIRECTOR, CAREER SERVICES
EASTERN MENNONITE UNIVERSITY
HARRISONBURG, VA 22801

*9815
LEONARD JONES, JR.
DIRECTOR, TESTING SVCS
HAMPTON UNIVERSITY
WIGWAM BUILDING - RM 200
HAMPTON, VA 23668
804-727-5000

*9307
ELIZA C. SMITH
DEANS OFFICE
J. SARGEANT REYNOLDS COMMUNITY COLLEGE
PO BOX 85622
RICHMOND, VA 23285-5622
804-786-3470

*8622
LEONARD PARKER
DIR, EXPERIENTIAL LEARNING
LIBERTY UNIVERSITY
BOX 11803
LYNCHBURG, VA 24506
804-582-7709

*8002
PAULA DEAN
COORDINATOR, LEARNING ASSISTANCE CENTER
LORD FAIRFAX COMMUNITY COLLEGE
PO BOX 47
173 SKIRMISHER LANE
MIDDLETOWN, VA 22645
540-869-1120

9214
PRICILLA MOORELOCK
PROGRAM COORDINATOR
CENTER FOR CONTINUING ED.
MARY WASHINGTON COLLEGE
1301 COLLEGE AVE
FREDRICKSBURG, VA 22401
703-899-4100

8033
ART CAVANAUGH
DEAN OF RECORDS & REGISTRAR
MARYMOUNT UNIVERSITY
2807 NORTH GLEBE RD
ARLINGTON, VA 22207-4299
703-284-1538

9240
RON SMITH
DIRECTOR OF ADMISSIONS
NATIONAL BUSINESS COLLEGE
1813 EAST MAIN ST
SALEM, VA 24153
703-986-1800

8794
CAROL MICHELSON
ACADEMIC SERVICES COUNSELOR
NATIONAL-LOUIS UNIVERSITY
8000 WEST PARK DR, SUITE 125
McLEAN, VA 22102
703-749-3000

8960
MARGARET T. CHRISLEY
COORD, ADMISSIONS AND RECORDS
NEW RIVER COMMUNITY COLLEGE
DRAWER 1127
DUBLIN, VA 24084
703-674-3600

9257
MAX BASSETT
DEAN OF ACAD. & STUDENT SVCS
NORTHERN VIRGINIA COMMUNITY COLLEGE
4001 WAKEFIELD CHAPEL RD
ANNANDALE, VA 22003
703-323-3000

*9265
SHARON JOYNER
ASSISTANT DIRECTOR OF TESTING
ACADEMIC SKILLS CENTER
OLD DOMINION UNIVERSITY
1500 WEST 48th ST
NORFOLK, VA 23529
757-683-3503

9883
CHRISTOPHER B. KNAUER
ACTING REGISTRAR
RADFORD UNIVERSITY
PO BOX 6904
RADFORD, VA 24042
703-831-5271

This directory lists colleges & universities that award credit for passing scores on DSSTs.

*** INSTITUTIONS THAT ADMINISTER DSSTs**

**** Some schools may restrict testing to students admitted to or enrolled in the institution or offer a limited testing schedule.**

Students should contact the test center to determine current policies and schedules for DANTES testing.

*9910
JOHN T. THIOS
DIRECTOR OF COUNSELING
RICHARD BLAND COLLEGE OF THE COLLEGE OF WILLIAM & MARY
11301 JOHNSON ROAD
PETERSBURG, VA 23805
804-862-6224

8382
CHARLES BEVINS
DIRECTOR
SAINT LEO COLLEGE
PO BOX 4326
FT EUSTIS, VA 23604
757-887-0655

9394
CAROL CORRIGAN
DIRECTOR, FT. LEE
SAINT LEO COLLEGE
PO BOX 5220
BLDG 8035 BISHOP LOOP
FT. LEE, VA 23801-0220
804-861-9634

9404
JON CARLETON
DIRECTOR, LANGLEY AFB
SAINT LEO COLLEGE
BASE EDUCATION CENTER, BLDG #1027
PO BOX 65629
LANGLEY AFB, VA 23665
757-766-1812

9397
EARNEST J. EVERETT
ASST DIRECTOR, NAVY CTR S. HAMPTON RDS
SAINT LEO COLLEGE
1328 8TH ST, SUITE 100
BLDG #3005, NAVPHIBASE
NORFOLK, VA 23521
757-464-6449

*9817
PAUL WILEY
REGISTRAR
SHENANDOAH UNIVERSITY
1460 COLLEGE DR
WINCHESTER, VA 22601
703-665-4506

8356
DONALD H. SMITH
DEAN OF INSTRUCTION
SOUTHWEST VIRGINIA COMMUNITY COLLEGE
BOX SCC
RICHLANDS, VA 24641
703-964-2555

8456
NATALIE MARSH
SR. ACADEMIC OFFICER
STRAYER COLLEGE
COMMAND EDUCATION
3098 RANGE RD, SUITE 102
QUANTICO, VA 22134
703-640-9926

9340
E. ROY BUDD
DIRECTOR, CENTER FOR BUSINESS/COMMUNITY SERVICES
THOMAS NELSON COMMUNITY COLLEGE
99 THOMAS NELSON DR
HAMPTON, VA 23670
804-825-2936

*9341
HAROLEEN RAY
COORDINATOR OF ENROLLMENT AND SERVICE
TIDEWATER COMMUNITY COLLEGE
7000 COLLEGE DR - STATE RT 135
PORTSMOUTH, VA 23703
757-822-2124

*8273
TERRI A. MILLER
ASSISTANT DIRECTOR
TIDEWATER TECH
616 DENGIGH BLVD
NEWPORT NEWS, VA 23602

8780
DIANNE L. BARRON
PROGRAM COORDINATOR
TROY STATE UNIVERSITY
PO BOX 1032, BLDG 219
FT. MYER, VA 22211

*8383
JERRY MUSICK
TESTING COORDINATOR
VIRGINIA HIGHLANDS COMMUNITY COLLEGE
PO BOX 828
ABINGDON, VA 24212
540-466-3444

WASHINGTON

*9063
CANDIS LACHER
DIRECTOR OF ADMISSIONS
BIG BEND COMMUNITY COLLEGE
7662 CHANUTE ST
MOSES LAKE, WA 98837
509-762-6228

*9024
MARY BELKNAP
REGISTRAR
CITY UNIVERSITY
335 - 116th AVE, SE
BELLEVUE, WA 98004
206-643-2000

8661
LINDA CALVERT
COORDINATOR OF ADMISSIONS
CLARK COLLEGE
1800 EAST McLOUGHLIN BLVD
VANCOUVER, WA 98663
206-694-6521

*9161
JAMES T. MANSFIELD
DIRECTOR OF ADMISSIONS
GONZAGA UNIVERSITY
SPOKANE, WA 99258
509-328-4220

9123
MARGO PERROTTI
REGISTRAR
HERITAGE COLLEGE
3240 FT. RD
TOPPENISH, WA 98948
509-865-2244

*8052
CAROL RICHARDS
ASST. TO REGISTRAR
NORTHWEST COLLEGE
PO BOX 579
KIRKLAND, WA 98083
206-889-5230

*9922
GARY MINETTI
COUNSELING & TESTING
PACIFIC LUTHERAN UNIVERSITY
TACOMA, WA 98447
206-531-6900

*8393
WENDY SHEA
COORDINATOR OF TESTING
PENINSULA COLLEGE
1502 EAST LAURIDSEN BLVD
PORT ANGELES, WA 98362
360-417-6346

8670
EVIE CLOSSON
EVALUATOR
PIERCE COLLEGE
9401 FARWEST DR, SW
TACOMA, WA 98498
206-964-6500

9294
LINDA FLETCHER
DIRECTOR
ST. MARTIN'S COLLEGE
HQ, I CORPS AND FORT LEWIS
ATTN: AFZH-PAE, MS 20g
BOX 339500
FORT LEWIS, WA 98433-9500
206-964-4688

8908
BOBBE MILLER-MURRAY
ASST. REGISTRAR
SEATTLE UNIVERSITY
900 BROADWAY AT MADISON
SEATTLE, WA 98122
206-296-5856

*9310
BRITTON SPRAGUE
DEAN
SKAGIT VALLEY COLLEGE
2405 COLLEGE WAY
MOUNT VERNON, WA 98273
206-428-1261

8660
DWIGHT LONG
PROGRAM MANAGER
SOUTH PUGET SOUND COMMUNITY COLLEGE
2011 MOTTMAN RD, SW
OLYMPIA, WA 98512
360-754-7711

This directory lists colleges & universities that award credit for passing scores on DSSTs.

* INSTITUTIONS THAT ADMINISTER DSSTs

**** Some schools may restrict testing to students admitted to or enrolled in the institution or offer a limited testing schedule.**

Students should contact the test center to determine current policies and schedules for DANTES testing.

8361
MS. FRAN M. MINIKEN
ASST. COORDINATOR
SOUTHERN ILLINOIS UNIVERSITY/CARBONDALE
62 MSS/DPE 132 THIRD ST
McCHORD AFB, WA 98438-5975
206-582-6561

9336
PHIL GRIFFIN
CREDENTIALS EDUCATOR
TACOMA COMMUNITY COLLEGE
6501 SOUTH 19th ST
TACOMA, WA 98466
206-566-5327

*9834
CELIA GILGER
OFFICE OF EDUCATIONAL ASSESSMENT
UNIVERSITY OF WASHINGTON
SEATTLE, WA 98195
206-543-1170

8748
DALE VISGER, CHAIRMAN
IND TECH DEPT
WALLA WALLA COLLEGE
204 SOUTH COLLEGE AVE
COLLEGE PLACE, WA 99324

*8559
JERRY NELSON
TESTING CENTER
WHATCOM COMMUNITY COLLEGE
237 WEST KELLOGG RD
BELLINGHAM, WA 98226
206-676-2170

9396
JANE T. GRUBENHOFF
CREDENTIALS EVALUATOR
YAKIMA VALLEY COMMUNITY COLLEGE
16th & NOB HILL BLVD
YAKIMA, WA 98907
509-575-2372

WEST VIRGINIA

*9090
KAREN CARTER-HARVEY
SAELL SENIOR STUDENT OUTREACH COORDINATOR
COLLEGE OF WEST VIRGINIA
PO BOX AG
609 SOUTH KANAWHA STREET
BECKLEY, WV 25802-2830
304-253-7351 EXT 1366

9117
LAWRENCE DICKEN
DIRECTOR OF ADMISSIONS
CONCORD COLLEGE
ATHENS, WV 24712
304-384-3115

9306
DAVID ELDRIDGE
REGISTRAR
SHEPHERD COLLEGE
SHEPHERDSTOWN, WV 25443
304-876-5463

*8915
ROBERT P. SCHOLL
REGISTRAR/ADMISSIONS
WEST VIRGINIA INSTITUTE OF TECHNOLOGY
210 OLD MAIN
MONTGOMERY, WV 25136
304-442-3071

*9847
DR. JOHN FULLER
DIRECTOR OF REGISTR. & RECORDS
WEST VIRGINIA STATE COLLEGE
P. O. BOX 1000
INSTITUTE, WV 25112
304-766-3144

8970
ELLIE SUMMERS
SUPERVISOR/TRANSFER ADMISSION
WEST VIRGINIA UNIVERSITY
PO BOX 6009
MORGANTOWN, WV 26506
1-800-344-9881

*9440
JANET ROSKOVICH
REGISTRAR
WHEELING JESUIT UNIVERSITY
316 WASHINGTON AVENUE
WHEELING, WV 26003-6295
(800) 873-7665

WISCONSIN

9987
MARK REINHOLTZ
DIRECTOR, CREDIT FOR PRIOR LEARNING
CARDINAL STRITCH UNIVERSITY
6801 NORTH YATES RD
MILWAUKEE, WI 53217
414-410-4092

9849
GREGORY SCHNEIDER
DIRECTOR OF COUNSELING
CARROLL COLLEGE
100 NORTH EAST AVE
WAUKESHA, WI 53186
414-547-1211

9989
RUTH JOHNSON
REGISTRAR
CARTHAGE COLLEGE
KENOSHA, WI 53141
414-551-8500

*8281
DONELL M. DELVEAUX
CENTER DIRECTOR
CONCORDIA UNIVERSITY
1150 SPRINGHURST DR, SUITE 101
GREEN BAY, WI 54304
920-498-2251

8852
REGISTRAR
CONCORDIA UNIVERSITY OF WISCONSIN
12800 NORTH LAKE SHORE DR
MEQUON, WI 53092
414-243-5700

*8255
DEBORAH DAVIDSON
DIRECTOR, ADULT LEARNING CTR
CONCORDIA UNIVERSITY OF WISCONSIN
7500 GREEN BAY RD
KENOSHA, WI 53142
414-697-8260

*9624
ERIC GROSSO
ADMISSIONS/CUW-MADISON
CONCORDIA UNIVERSITY WISCONSIN
2909 LANDMARK PLACE, STE 101
MADISON, WI 53713
608-277-7900

*8300
JOAN SCHILLING
COORDINATOR FOR PRIOR LEARNING
EDGEWOOD COLLEGE
855 WOODROW ST
MADISON, WI 53711
608-257-4861

9850
F. JAMES BENIESSE
DIRECTOR, PROGRAM WISE
LAKELAND COLLEGE
PO BOX 359, COUNTRY TRUNK M
SHEBOYGAN, WI 53082
414-565-1202

*8285
MICHAEL SCHULTE
MANAGER VOCATIONAL ASSESSMENT
LAKESHORE TECHNICAL COLLEGE
1290 NORTH AVE
CLEVELAND, WI 53015
920-458-4183, X224

8989
DAVID L. HERSHBERGER
REGISTRAR
MARANATHA BAPTIST BIBLE COLLEGE
745 WEST MAIN STREET
WATERTOWN, WI 53094
414-261-9300

8685
ANTHONY D. TORTORELLA
UNIVERSITY REGISTRAR
MARQUETTE UNIVERSITY
PO BOX 1881
MILWAUKEE, WI 53201-1881
414-288-7302

9852
MARLENE PENFIELD
OUTREACH ADMISSIONS COORDINATOR
MOUNT SENARIO COLLEGE
1500 COLLEGE AVE WEST
LADYSMITH, WI 54848
715-532-5511

8572
EDWARD ZAHN
REGISTRAR
NORTH CENTRAL TECHNICAL COLLEGE
1000 CAMPUS DR
WAUSAU, WI 54401
715-675-3331

***8914**
JACKY THOMSON
FINANCIAL AID COORDINATOR
OTTAWA UNIVERSITY - MILWAUKEE
300 NORTH CORP. DR, SUITE 110
BROOKFIELD, WI 53045
414-879-0200

9856
MILLARD STOREY
DIRECTOR OF ADMISSIONS
UNIVERSITY OF WISCONSIN
750 UNIVERSITY AVE
MADISON, WI 53706
608-262-3961

9997
SUSAN FELBER
ASST DIRECTOR OF ADMISSIONS
WW - EAU CLAIRE
112 SCHOFIELD HALL
EAU CLAIRE, WI 54702-4004
715-836-5415

***9372**
TIMOTHY HOYT
DIR. RESEARCH & TESTING SVCS
UNIVERSITY OF WISCONSIN
800 ALGOMA BLVD
OSHKOSH, WI 54901
414-424-1401

9998
MARY POWER
COORD., ADVISING AND TESTING
UNIVERSITY OF WISCONSIN-PARKSIDE
900 WOOD ROAD, BOX 2000
KENOSHA, WI 53141
414-595-2321/2219

9373
RICHARD SCHUMACHER
DEAN OF ADMISSIONS
UNIVERSITY OF WISCONSIN
1 UNIVERSITY PLAZA
PLATTEVILLE, WI 53818
608-342-1491

***9374**
SHARON LAVINE
COORDINATOR, ASSESSEMNT OF PRIOR LEARNING
UNIVERSITY OF WISCONSIN
1800 GRAND AVE
SUPERIOR, WI 54880
715-394-8499

9375
TORI McGUIRE
ASST DIRECTOR OF ADMISSIONS
UNIVERSITY OF WISCONSIN
148 BAKER HALL
WHITEWATER, WI 53190
414-472-1512

***9554**
VICKI FAORO ANDERSON
DIRECTOR
UPPER IOWA UNIVERSITY - WAUSAU CENTER
1000 CAMPUS DRIVE
ROOM 112A
WAUSAU, WI 54401
727-864-8163

***8256**
LINDA HAINES
DIRECTOR
UPPER IOWA UNIVERSITY
MILWAUKEE WEST CENTER
6610 W GREENFIELD AVE
WEST ALLIS, WI 53214

***8532**
TODD ALLGOOD
DIRECTOR EVALUATION AND PLACEMENT
NICOLET TECHNICAL COLLEGE
PO BOX 518
RHINELANDER, WI 54501

WYOMING

***8779**
ALICE NICOL
COUNSELOR
CENTRAL WYOMING COLLEGE
2660 PECK AVE
RIVERTON, WY 82501
307-855-2175

***9457**
MARY GRIDLEY
DIRECTOR
CENTRAL WYOMING COLLEGE - TETON CAMPUS
PO BOX 568
220 SOUTH GLENWOOD
JACKSON, WY 83001
307-733-7425

8418
ADELL VAN PATTEN-GORNY
DIRECTOR
CHAPMAN COLLEGE, WARREN A.C.
90 MSSQ/MSE
FE WARREN AFB, WY 82005
307-632-9927

***8607**
CONNIE WOEHL
CAMPUS DIRECTOR
EASTERN WYOMING COLLEGE
203 N. SIXTH ST.
DOUGLAS, WY 82633
307-358-5622

9195
EDMOND BOENISCH
DEAN OF STUDENTS
LARAMIE COUNTY COMMUNITY COLLEGE
1400 EAST COLLEGE DR
CHEYENNE, WY 82007
307-778-1213

***8276**
SHARON GUERNEY
DIRECTOR, WYOMING PROGRAMS
REGIS UNIVERSITY
1400 EAST COLLEGE DR
CHEYENNE, WY 82007-3299
307-778-1282

OVERSEAS INSTITUTIONS

***8165**
EILEEN LEE
TESTING SUPERVISOR
ANGLO-CHINESE JUNIOR COLLEGE
19 TOH TUCK RD #02-03
SINGAPORE 596683
65-4632721

***8721**
A. A. AJARIMAH
SUPERVISOR, TESTING & EVALUATION UNIT
ARAMCO
TRAINING/PROGRAM DEVELOPMENT
T-890
DHAHRAN
SAUDI ARABIA 31311

***9955**
SHERIDAN TALBOT
DIRECTOR, STUDENT SERVICES
BERMUDA COLLEGE
2nd LEVEL, LIBRARY BLDG
RM L 150 STONINGTON CAMPUS
PAGET, BERMUDA PG 04
441-236-9000 EXT 290

***8127**
ANNE BARRADEEN
EXTERNAL EXAM SUPERVISOR
BREMER INSTITUTE OF TAFE
MARY STREET
BUNDAMBA Q 4304
AUSTRALIA
07-3817-3107

***8710**
JOHN HOBBS
PRESIDENT
BUSINESS SCHOOL OF LAUSANNE
PO BOX 2290 AVE EO. DAPPLES 38
SWITZERLAND 1066

***8849**
JAMES BUTLER
DIRECTOR
CENTER FOR AMERICAN EDUCATION
PO BOX 12867, DUBAI
UNITED ARAB EMIRATES
971-4-627755

9952
JOHNELLE WELSH
ASSOCIATE DEAN
CENTRAL TEXAS COLLEGE (KOREA)
PFEC CAMP MARKET
BUILDING P-1530
UNIT #15559
APO, AP 96283

8390
PETER HERSHFIELD
AREA DIRECTOR
CENTRAL TEXAS COLLEGE (PANAMA)
UNIT 1319
APO, AA 34004
507-287-3773

*** INSTITUTIONS THAT ADMINISTER DSSTs**

**** Some schools may restrict testing to students admitted to or enrolled in the institution or offer a limited testing schedule.**

Students should contact the test center to determine current policies and schedules for DANTES testing.

9957
ATTN: TRANSCRIPTS
CITY COLLEGES OF CHICAGO
UNIT 29651
APO, AE 09096
011-49-6134-26633

9326
STEVE JOHNSON
DIRECTOR
EMBRY-RIDDLE AERONAUTICAL UNIVERSITY
EUROPEAN DIVISION
CMR 429
APO AE 09095

***8976**
DAVID G. WATT
VP, ACADEMIC AFFAIRS
GUAM COMMUNITY COLLEGE
PO BOX 23069 GMF
BARRIGADA, GAUM 96921
671-735-5527

***8962**
BEATRICE RIVERA
DIRECTOR, COUNSELING PROGRAM
INTER AMERICAN UNIVERSITY/PUERTO RICO
RECINTO METROPOLITANO - PO BOX 19293
SAN JUAN, PR 00936
787-759-6327

***8044**
MARTHA HARRISON
COORDINATOR, CENTER FOR DISTANCE LEARNING
KING FAISAL HOSPITAL/RESEARCH CENTRE
PO BOX 3354
NURSING ED & RES MBC #63
RIYADH, SAUDI ARABIA 11211

***8706**
HOWARD BROWN
DIRECTOR OF ADULT EDUCATION
KWAJALEIN SCHOOL SYSTEM
PO BOX 54
APO, AP 96555
805-355-1078

***8767**
ELIZABETH LANGSTEN
COORDINATOR
MALAYSIAN AMERICAN COMMSN. ON EDUCATIONAL. EXCHANGE.
(MACEE) PENANG
FIRST FLOOR, KOLEJ DAMANSARA UTAMA PENANG
32 JALAN ANSON
10400 PENANG
MALAYSIA
PH: 604-226-4924

***8869**
MEENA PONNUSAMY
TESTING COORDINATOR
MALAYSIAN AMERICAN COMMSN. ED. EXCH.
181 JALAN TUN RAZAK
50400 KUALA LUMPUR
MALAYSIA 50450

***8395**
LARRY D. McCULLOUGH
PRESIDENT
NORTHWEST BAPTIST COLLEGE
22606-76 AVE
PO BOX 790
LANGLEY, B.C. V34A8B8
604-888-3310

***8895**
TSVI BISK
DIRECTOR
S.T.E.P. INSTITUTE
6 REINES AVE
TEL-AVIV
ISRAEL 64381
972-3-5278490

9939
DR. PAUL BREWER
DIRECTOR, STUDENT SERVICES
UNIVERSITY OF MARYLAND
EUROPEAN DIVISION
UNIT 29216
APO, AE 09102

***9940**
RICK HESS
DIRECTOR, STUDENT SERVICES
UNIVERSITY OF MARYLAND
ASIAN DIVISION, BOX 0100
APO, AP 96328-0100

***8029**
PHILLICIA SPRAUVE
TESTING OFFICE
UNIVERSITY OF THE VIRGIN ISLANDS
2 JOHN BREWER'S BAY
ST. THOMAS, VI 00802
809-776-9200

***8907**
DR. ROBERT A. SPENCER
DIRECTOR
WEBSTER UNIVERSITY - GENEVA
15, ROUTE DE COLLEX
1293 BELLEVUE
GENEVA, SWITZERLAND
011-41-22-774-2452

***8766**
M.C. EGURIDU
TEST DEVELOPMENT DIVISION
WEST AFRICAN EXAMINATION COUNCIL
PMB 1076
YABA LAGOS, NIGERIA

***8866**
J.Y. NEWLANDS
SENIOR DEPUTY REGISTRAR
WEST AFRICAN EXAMINATIONS COUNCIL
PO BOX 917
ACCRA, GHANA

DANTES

Subject Standardized Tests

Fact Sheet

Study Guide

ASTRONOMY
SF 500
(REVISED)

TEST INFORMATION

This test was developed to enable schools to award credit to students for knowledge equivalent to that which is learned by students taking the course. The test contains 87 items. The American Council on Education recommends that credit be awarded for a minimum of 42 correct responses, which is a standard score of 48. This minimum credit-awarding score is equal to the mean score of students in the norming sample who received a grade of C in the course. However, some schools set their own standards for awarding credit and may require a higher score. Students should obtain this information from the institution where they expect to receive credit.

CONTENT

The following topics, which are commonly taught in courses on this subject, are covered by this examination.

	Approximate Percent
I. History of Astronomy	9%
A. Nature of science	
B. How scientists think and work	
II. Celestial Mechanics, Including Gravitation and Relativity	5%
III. Celestial Systems	13%
A. Earth and the sky	
B. Earth and the Moon	
C. Time and the calendar	
IV. Astronomical Instruments	12%
A. Measurement and analysis of starlight	
B. The electromagnetic spectrum	
V. The Solar System	19%
A. Contents, form, and motions	
B. Evolution	
VI. The Sun and Stars: Nature and Evolution	17%
VII. Our Galaxy: Contents and Structure	7%
VIII. The Universe: Contents, Structure, and Evolution	10%
IX. Determining Astronomical Distances	5%
X. Life in the Universe	3%

Questions on the test require candidates to demonstrate the following abilities. Some questions may require more than one of the abilities.

- Knowledge of basic facts and terms (about 45 - 50% of the examination)
- Understanding of concepts and principles (about 40 - 45% of the examination)
- Ability to apply knowledge to specific cases or issues (about 5 - 10% of the examination)

SAMPLE QUESTIONS

1. In the northern hemisphere, the vernal equinox is the position occupied by the Sun on the first day of

    (A) summer
    (B) fall
    (C) spring
    (D) winter

2. Which of the following is the best illustration of Newton's third law?

    (A) A skater coasting across the ice
    (B) The spinning of a top
    (C) The swinging of a pendulum
    (D) The recoil of a shotgun

3. The energy in the interior of a white dwarf is transported outward in the same fashion as the energy is transmitted

    (A) in an airplane shockwave
    (B) from an electric oven
    (C) from a hot-air furnace
    (D) from tip to handle of a hot poker

4. The most important advantage of a large telescope aperture is that it

    (A) allows a large amount of radiation to be collected
    (B) gives a higher magnification of the objects observed
    (C) is less affected by the trembling of the Earth's atmosphere
    (D) produces a larger diffraction ring when distant stars are observed

5. Which of the following statements is true about the steady-state cosmology?

    (A) It explains the isotropic nature of the remnant radiation from a giant fireball.
    (B) It appears to violate the law of conservation of matter in empty space.
    (C) It predicts a negative value for the Hubble Constant.
    (D) It explains the galactic red shifts as gravitational effects.

6. Where is the Moon when there are spring tides on Earth?

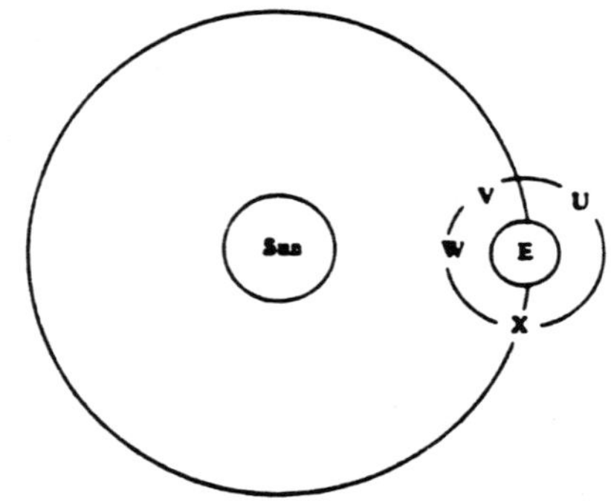

E = Earth
U, V, W, X = Positions of the Moon

    (A) U
    (B) V
    (C) W
    (D) X

7. The bending of rays of light as they pass from one transparent medium into another is called

    (A) reflection
    (B) diffraction
    (C) dispersion
    (D) refraction

8. Which of the following planets has been observed to have extensive Van Allen belts similar to those of Earth?

    (A) Mercury
    (B) Mars
    (C) Jupiter
    (D) Venus

9. The fact that most stars observed are on the Main Sequence implies that a star spends the greatest proportion of its lifetime

    (A) contracting to reach the Main Sequence
    (B) as a giant off the Main Sequence
    (C) expanding to reach the Main Sequence
    (D) on the Main Sequence

10. At the present time in the Sun's lifetime, the major source of the Sun's energy lies in

(A) electron-proton collisions
(B) gravitational contraction
(C) nuclear fusion
(D) matter-antimatter annihilation

11. The Universe as we know it began its existence as a hot, dense cloud of matter and radiation approximately how many years ago?

(A) 5 billion
(B) 15 billion
(C) 50 billion
(D) 100 billion

REFERENCE PUBLICATIONS

These are some of the books that may be helpful in preparing for the test, although any standard text in the subject is probably appropriate.

Abell, George *Exploration of the Universe.* 5th ed. New York: CBS/Saunders College Publishing, 1987.

Jastrow, Robert, and Malcolm H. Thompson. *Astronomy: Fundamentals and Frontiers.* 4th ed. New York: John Wiley and Sons, Inc., 1984.

Kaufmann, William J. *Discovering the Universe. 1st* ed. New York: Freeman, 1987.

Pasachoff, Jay M. *Astronomy: From Earth to the Universe.* 3d ed. Philadelphia: Saunders College Publishing, 1987.

Seeds, Michael. *Horizons: Exploring the Universe.* 3d ed. Belmont, CA: Wadsworth, 1989.

Zeilik, Michael. *Astronomy: The Evolving Universe.* 5th ed. New York: John Wiley and Sons, Inc., 1982.

CREDIT RECOMMENDATIONS

The Center for Adult teaming and Educational Credentials of the American Council on Education (ACE) has reviewed and evaluated the DANTES test development process and has made the following recommendations:

Area or Course
Equivalent: Astronomy
Level: Lower-level Baccalaureate
Amount: Three (3) semester hours
Minimum Credit-awarding
Score: Scaled score of 48
(42 correct responses)
Source: ACE Commission on Educational Credit and Credentials

TECHNICAL INFORMATION

Twenty-eight (28) colleges and universities participated in the norming during the fall of 1990. These schools ranged from large state universities to small private schools. Geographic distribution was wide, with schools from all parts of the United States involved. A total of 746 students who were completing or had recently completed a course in this subject took part in the norming.

In 1995, the test was reviewed by faculty members who currently teach the course. On their recommendation, the test was revised slightly. The current form of the test was equated to the previous form.

Raw scores on the current form are converted to scaled scores by applying the formula S=AX + B where S= the scaled score, X= the raw score, the slope (A) = .7700, and the intercept (B) = 15.4660. The reporting scale has a mean of 50 and a standard deviation of 10.

NORMING STATISTICS
(based on 745 students)

Number of items	87
Mean (Raw Score)	44.9
Standard deviation (Raw Score)	13.0
KR-20 Reliability	.90
Standard Error of measurement	
(Raw Score)	4.1
(Scaled Score)	3.2

Average Grade/Test Score Correlation, based on a sample of 636 students at 24 institutions (an average of within-school correlations, weighted by the number of students tested in each school): .64

Average scaled score for students whose teachers gave them a grade of A for the course: 58

Average scaled score for students whose teachers gave them a grade of B for the course: 52

Average scaled score for students whose teachers gave them a grade of C for the course: 48

Colleges and universities that would like to have review copies of tests, additional information about the national norming, or assistance in local norming or score validation studies should write to: DANTES Program, Mail Stop 31-X, Educational Testing Service, Princeton, New Jersey 08541-0001.

Correct responses to sample questions: 1.C; 2.D; 3.D; 4.A; 5.B; 6.C; 7.D; 8.C; 9.D; 10.C; 11.B.

HOW TO TAKE A TEST

You have studied hard, long, and conscientiously.

With your official admission card in hand, and your heart pounding, you have been admitted to the examination room.

You note that there are several hundred other applicants in the examination room waiting to take the same test.

They all appear to be equally well prepared.

You know that nothing but your best effort will suffice. The "moment of truth" is at hand: you now have to demonstrate objectively, in writing, your knowledge of content and your understanding of subject matter.

You are fighting the most important battle of your life — to pass and/or score high on an examination which will determine your career and provide the economic basis for your livelihood.

What extra, special things should you know and should you do in taking the examination?

BEFORE THE TEST

YOUR PHYSICAL CONDITION IS IMPORTANT

If you are not well, you can't do your best work on tests. If you are half asleep, you can't do your best either. Here are some tips:

1. Get about the same amount of sleep you usually get. Don't stay up all night before the test, either partying or worrying — DON'T DO IT.
2. If you wear glasses, be sure to wear them when you go to take the test. This goes for hearing aids, too.
3. If you have any physical problems that may keep you from doing your best, be sure to tell the person giving the test. If you are sick or in poor health, you really cannot do your best on any test. You can always come back and take the test some other time.

AT THE TEST

EXAMINATION TECHNIQUES

1. Read the *general* instructions carefully. These are usually printed on the first page of the examination booklet. As a rule, these instructions refer to the timing of the examination; the fact that you should not start work until the signal and must stop work at a signal, etc. If there are any *special* instructions, such as a choice of questions to be answered, make sure that you note this instruction carefully.

2. When you are ready to start work on the examination, that is as soon as the signal has been given, read the instructions to each question booklet, underline any key words or phrases, such as *least*, *best*, *outline*, *describe*, and the like. In this way you will tend to answer as requested rather than discover on reviewing your paper that you *listed without describing*, that you selected the *worst* choice rather than the *best* choice, etc.

3. If the examination is of the objective or so-called multiple-choice type, that is, each question will also give a series of possible answers: A, B, C, or D, and you are called upon to select the best answer and write the letter next to that answer on your answer paper, it is advisable to start answering each question in turn. There may be anywhere from 50 to 100 such questions in the three or four hours allotted and you can see how much time would be taken if you read through all the questions before beginning to answer any. Furthermore, if you come across a question or a group of questions which you know would be difficult to answer, it would undoubtedly affect your handling of all the other questions.

4. If the examination is of the essay-type and contains but a few questions, it is a moot point as to whether you should read all the questions before starting to answer any one. Of course if you are given a choice, say five out of seven and the like, then it is essential to read all the questions so you can eliminate the two which are most difficult. If, however, you are asked to answer all the questions, there may be danger in trying to answer the easiest one first because you may find that you will spend too much time on it. The best technique is to answer the first question, then proceed to the second, etc.

5. Time your answers. Before the examination begins, write down the time it started, then add the time allowed for the examination and write down the time it must be completed, then divide the time available somewhat as follows:

    a. If 3 ½ hours are allowed, that would be 210 minutes. If you have 80 objective-type questions, that would be an average of about 2 ½ minutes per question. Allow yourself no more than 2 minutes per question, or a total of 160 minutes, which will permit about 50 minutes to review.

    b. If for the time allotment of 210 minutes, there are 7 essay questions to answer, that would average about 30 minutes a question. Give yourself only 25 minutes per question so that you have about 35 minutes to review.

6. The most important instruction is *to read each question* and make sure you know what is wanted. The second most important instruction is to *time yourself properly* so that you answer every question. The third most important instruction is to *answer every question*. Guess if you have to but include something for each question, Remember that you will receive no credit for a blank and will probably receive some credit if you write something in answer to an essay question. If you guess a letter, say "B" for a multiple-choice question, you may have guessed right. If you leave a blank as the answer to a multiple-choice question, the examiners may respect your feelings but it will not add a point to your score. Some exams may penalize you for wrong answers, so in such cases *only*, you may not want to guess unless you have some basis for your answer.

7. Suggestions

   a. Objective-Type Questions

      (1) Examine the question booklet for proper sequence of pages and questions.

      (2) Read all instructions carefully.

      (3) Skip any question which seems too difficult; return to it after all other questions have been answered.

      (4) Apportion your time properly; do not spend too much time on any single question or group of questions.

      (5) Note and underline key words—*all*, *most*, *fewest*, *least*, *best*, *worst*, *same*, *opposite*.

      (6) Pay particular attention to negatives.

      (7) Note unusual option, e.g., unduly long, short, complex, different or similar in content to the body of the question.

      (8) Observe the use of "hedging" words—*probably*, *may*, *most likely*, *etc*.

      (9) Make sure that your answer is put next to the same number as the question.

      (10) Do not second guess unless you have good reason to believe the second answer is definitely more correct.

      (11) Cross out original answer if you decide another answer is more accurate; do not erase, *until* you are ready to hand your paper in.

      (12) Answer all questions; guess unless instructed otherwise.

      (13) Leave time for review.

b. Essay-Type Questions

(1) Read each question carefully.

(2) Determine exactly what is wanted. Underline key words or phrases.

(3) Decide on outline or paragraph answer.

(4) Include many different points and elements unless asked to develop any one or two points or elements.

(5) Show impartiality by giving pros and cons unless directed to select one side only.

(6) Make and write down any assumptions you find necessary to answer the question.

(7) Watch your English, grammar, punctuation, choice of words.

(8) Time your answers; don't crowd material.

8. Answering the Essay Question

Most essay questions can be answered by framing the specific response around several key words or ideas. Here are a few such key words or ideas:

M's: manpower, materials, methods, money, management

P's: purpose, program, policy, plan, procedure, practice, problems, pitfalls, personnel, public relations

a. Six basic steps in handling problems:

(1) preliminary plan and background development

(2) collect information, data and facts

(3) analyze and interpret information, data and facts

(4) analyze and develop solutions as well as make recommendations

(5) prepare report and sell recommendations

(6) install recommendations and follow up effectiveness

b. Pitfalls to Avoid

(1) *Taking Things for Granted*
A statement of the situation does not necessarily imply that each of the elements is necessarily true; for example, a complaint may be invalid and biased so that all that can be taken for granted is that a complaint has been registered

(2) *Considering only one side of a situation*
Wherever possible, indicate several alternatives and then point out the reasons you selected the best one.

(3) *Failing to indicate follow up*
Whenever your answer indicates action on your part, make certain that you will take proper follow-up action to see how successful your recommendations, procedures, or actions turn out to be.

(4) *Taking too long in answering any single question*
Remember to time your answers properly.

EXAMINATION SECTION

ASTRONOMY
EXAMINATION SECTION
TEST 1

DIRECTIONS: Each question or incomplete statement is followed by several suggested answers or completions. Select the one that BEST answers the question or completes the statement. *PRINT THE LETTER OF THE CORRECT ANSWER IN THE SPACE AT THE RIGHT.*

1. In their studies of the heavens, the Babylonians tried to 1. ___
   A. predict the occurrence of "guest stars"
   B. make geometrical models of the motions of celestial objects
   C. predict the future positions of celestial objects
   D. make geometrical models of celestial objects

Questions 2-3.

DIRECTIONS: In questions 2 and 3, match the letter of the appropriate choice with the statement.

A. 365 1/4 days B. None of these
C. 364 1/4 days D. 366 1/4 days

2. The period of the Earth's revolution about the Sun. 2. ___

3. A possible synodic period of a superior planet. 3. ___

4. During the year the Sun appears to move with respect to the stars along 4. ___
   A. the ecliptic B. the celestial equator
   C. the horizon D. a deferent

5. For which of the following did the model of Eudoxus NOT account? 5. ___
   A. Retrogade motion
   B. Changes in brightness of the planets
   C. Inclination of planetary orbits
   D. Diurnal motion

6. The model of Eudoxus consisted of 6. ___
   A. epicycles and deferents
   B. eccentrics
   C. elliptical motions
   D. concentric spheres with various points of attachment

7. An essential assumption made by Eratosthenes when he measured the size of the Earth was that 7. ___
   A. the Sun was near by
   B. the Earth could be considered flat if one was concerned only with small areas
   C. the Earth was spherical
   D. Alexandria was generally north of Syene

8. The obtaining of a reasonable result for the ratio of the Earth-Sun distance to the Earth-Moon distance by Aristarchus must be considered fortuitous because 8. ___
   A. the Sun is further away than he thought
   B. the Moon's orbit is not circular
   C. the Moon's motion is not uniform
   D. all of the above

9. Ptolemy's model of the universe was based upon the assumption that 9. ___
   A. uniform motion in a straight line is the most natural motion for a celestial object
   B. the Earth is at the center of the orbits of the planets
   C. uniform motion in a circle is the most natural motion for a celestial object
   D. celestial objects are pulled into circular orbits by gravity

10. The sun moves with respect to the stars about 10. ___
    A. 24° per day B. 0° per day
    C. 1/2° per day D. 1° per day

11. To locate an object in space, one needs 11. ___
    A. one coordinate B. two coordinates
    C. three coordinates D. four coordinates

12. To locate an object on the celestial sphere, one needs 12. ___
    A. one coordinate B. two coordinates
    C. three coordinates D. four coordinates

Questions 13-16.
Directions: Match the letter of the CORRECT choice with questions 13 through 16.
   A. eastward along the celestial equator from the Vernal Equinox
   B. along the horizon from the north point through the east point
   C. along a vertical circle from the horizon
   D. along an hour circle from the celestial equator

13. Right ascension is measured. 13. ___

14. Declination is measured. 14. ___

15. Altitude is measured. 15. ___

16. Azimuth is measured. 16. ___

17. In the diagram which appears on the following page, of a part of the celestial sphere, what is the altitude of the star? The star is west of the meridian and the observer's latitude is 0°. 17. ___

A. 90° B. 45° C. 270° D. 0°

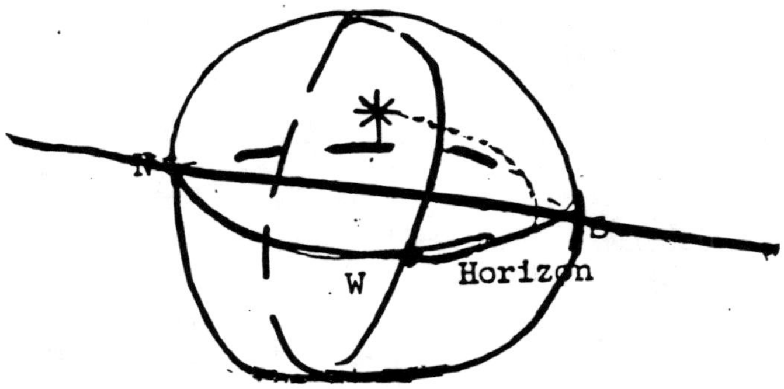

18. What is the declination of the star? 18. ___
A. 90° B. 45° C. 270° D. 0°

19. What is the sideral period of an inferior planet whose synodic period is 3/4 year? 19. ___
A. - 3 year B. -1/3 year C. 7/3 year D. 3/7 year

20. What is the synodic period of a superior planet whose sidereal period is 5 years? 20. ___
A. -5/4 year B. 5/4 year C. 4/5 year D. 5/6 year

21. The synodic period of a planet is the time it takes to 21. ___
A. revolve about the sun
B. revolve about the earth
C. move around the sky with respect to the stars
D. move around the sky with respect to the sun

22. The Copernican theory 22. ___
A. did away with the equant
B. put the sun at the center of the orbits of the planets
C. put the earth at the center of the orbits of the planets
D. did away with epicycles

23. The MAIN contribution of Tycho to astronomy was 23. ___
A. accurate observations of the positions of the planets
B. the discovery of a new star
C. an improved model of the solar system
D. the study of motion of a great comet

24. To determine the distance of an inferior planet from the sun in astronomical units,one needs to know 24. ___
A. the sideral period
B. the angle between the planet and the sun at the greatest elongation
C. either A or B
D. both A and B

25. What is the eccentricity of the orbit of a planet whose MAXIMUM and MINIMUM distances from the sun are 1 A.U. and 1/4 A.U.? 25. ___
A. 3/5 B. 5/4 C. 2/5 D. 5/8

TEST 2

DIRECTIONS: Each question or incomplete statement is followed by several suggested answers or completions. Select the one that BEST answers the question or completes the statement. *PRINT THE LETTER OF THE CORRECT ANSWER IN THE SPACE AT THE RIGHT.*

1. What is the MINIMUM distance from the sun of a planet whose orbit has an eccentricity of 1/2 and a semimajor axis of 2 A.U.? 1. ____
   A. 1 A.U. B. 2 A.U. C. 3 A.U. D. None of these

2. What would be the period of this planet? 2. ____
   A. 1 year B. $\sqrt[3]{4}$ years C. 8 years D. $\sqrt{8}$ years

3. What is the gravitational attraction between two people when 3. ____
   $m_1 = 100 \times 10^3$ gm, $m_2 = 60 \times 10^3$ gm,
   dist = 20 cm, $G = 7 \times 10^{-8}$ dynes cm^2 gm^{-2}?
   Consider that they act as point masses.
   A. $\frac{42}{2} \times 10^0$ dynes B. $\frac{42}{4} \times 10^{-1}$ dynes
   C. $\frac{42}{4} \times 10^3$ dynes D. $\frac{42}{4} \times 10^{19}$ dynes

4. How much would a 100-pound person weigh on Jupiter? The mass of Jupiter is about 300 times that of the earth, and its radius is about 11 times that of the earth. 4. ____
   A. $100 \times \frac{300}{(11)}$ lb. B. $100 \times \frac{300}{(11)^2}$ lb.
   C. $100 \times \frac{7 \times 10^{-8} \times 300}{(11)^2}$ lb. D. $100 \times \frac{7 \times 10^{-8} \times 300 \times 100}{(11)^2}$ lb.

5. What would be the acceleration of gravity on Mars? The mass of Mars is about 1/10 that of the earth, and the radius of Mars is about 1/2 that of the earth. 5. ____
   A. 2/5 B. 1/5 C. 50 D. None of these

6. A parabolic orbit is one in which the object 6. ____
   A. has a velocity equal to $\sqrt{2}$ times the circular velocity
   B. will escape from the object about which it is orbiting
   C. both A and B
   D. neither A or B

7. How long will it take a space probe to go from the earth to Jupiter on a minimum energy transfer orbit? The semimajor axis of Jupiter's orbit is about 5 A.U., and of the earth's orbit is 1 A.U. 7. ____
   A. $\frac{1}{2}(4^{3/2})$ B. $3^{3/2}$ C. $\frac{1}{2}(3^{3/2})$ D. $\frac{1}{2}(6^{3/2})$

8. A proof that the earth rotates is 8. ___
   A. the diurnal motion
   B. the change of directions of a Foucault pendulum
   C. the wind
   D. the seasons

9. The mass of the earth can be measured by 9. ___
   A. comparing the attraction between the earth and a given mass, M_A, to the attraction between another mass, M_B, and M_A
   B. multiplying the density of a small sample of the earth by the volume of the earth
   C. measuring the altitude of stars from different, known positions on the earth
   D. none of the above

10. The presently accepted age of the earth is about 10. ___
    A. 10 million years B. 1 billion years
    C. 4 billion years D. 10 billion years

11. At the North Pole, the stars appear to move 11. ___
    A. in circles around the north point on the horizon
    B. in circles around a point 45° between the north point on the horizon and the zenith
    C. in circles around the zenith
    D. in circles around the south point on the horizon

12. The speed at the equator due to the earth's rotation is 12. ___
    A. about 1000 miles/hour B. about 750 miles/hour
    C. about 0 miles/hour D. about 500 miles/hour

13. About what time of day or night does the moon rise when it is full? 13. ___
    A. midnight B. sunrise
    C. noon D. sunset

14. Which of the following can NEVER appear at opposition? 14. ___
    A. Saturn B. Moon C. Mars D. Sun

15. Suppose you were an astronaut in orbit about the earth and you wish to catch up with another space craft in the same orbit as you but some distance ahead. You should 15. ___
    A. fire rockets to increase your speed
    B. fire rockets to decrease your speed
    C. radio for help
    D. give up

16. It is December, and the moon is in its new phase. An observer on the earth is 16. ___
    A. in the umbra of the moon's shadow
    B. in the penumbra of the moon's shadow
    C. not in the moon's shadow, but in the earth's shadow
    D. not in either the moon's or the earth's shadows

17. It is December, and Mars and Jupiter are simultaneously at opposition. 17. ___
    A. Neither planet shows retrograde motion.
    B. Both planets show retrograde motion.
    C. Only one planet can show retrograde motion.
    D. Insufficient information to determine whether retrograde motion occurs

18. The sidereal period of Venus is 0.615 years. How long is the period between two successive superior conjunctions of Venus? 18. ___
    A. 0.615 years B. 1.000 years
    C. 1.598 years D. 2.628 years

19. Pluto's distance from the sun varies from 30 A.U. to 50 A.U. What is the APPROXIMATE eccentricity of its orbit? 19. ___
    A. .250 B. .400 C. .600 D. .666

20. A satellite in orbit is hit by a meteor of a thousandth its mass but equal momentum coming from the opposite direction. The meteor remains imbedded in the vehicle. The satellite 20. ___
    A. slows slightly
    B. is thrown out into space
    C. goes into a lower orbit
    D. drops straight down

21. An astronaut in orbit is taking a "space walk" outside of his spacecraft. He leaves a tool floating in space several hundred feet in front of the spacecraft (ahead of it in the same orbit with the same orbital velocity). 21. ___
    A. It remains at this same relative position in the spacecraft's orbit.
    B. It orbits the spacecraft at a distance of about several hundred feet from it.
    C. It slowly moves toward the spacecraft.
    D. It slowly moves away from the spacecraft.

22. Two stars, each having a mass equal to that of the sun, revolve about their barycenter at 1.0 A.U. from it. Their period of revolution is: 22. ___
    A. 0.5 sidereal years B. 1.0 sidereal years
    C. 2.0 sidereal years D. 4.0 sidereal years

23. Which of the following orbital parameters from the solar system is defined with respect to the earth's orbit? 23. ___
    A. Period B. Eccentricity C. Perihelion D. Inclination

24. Which of Newton's laws governs the path of an object in orbit about the sun? 24. ___
    A. First Law, momentum conserved
    B. Second Law, force = mass X acceleration
    C. Third Law, forces are mutual
    D. All of them

25. Eclipses of the sun can occur when the earth is APPROXIMATE-LY 25. ___
    A. at perihelion
    B. at the line of nodes
    C. at aphelion
    D. at any point in the earth's orbit

TEST 3

DIRECTIONS: Each question or incomplete statement is followed by several suggested answers or completions. Select the one that BEST answers the question or completes the statement. *PRINT THE LETTER OF THE CORRECT ANSWER IN THE SPACE AT THE RIGHT.*

1. The longitude 51° can also be expressed as 1. ___
   A. 51^h B. 12^h12^m C. $3^h1\text{-}1/2^m$ D. 3^h24^m

2. The non-uniformity of apparent solar time is caused by 2. ___
   A. eccentricity of the earth's orbit
   B. the inclination of the ecliptic to the celestial equator
   C. both A and B
   D. neither A or B

3. The rotation of the earth provides a uniform measure of the passage of time. 3. ___
   A. True B. False C. Relative D. Impossible to determine

4. The Gregorian calendar differs from the Julian calendar by
   A. omitting the 13th month in "full years"
   B. giving even-numbered months 30 days and odd-numbered months 31 days
   C. omitting 3 leap years in every 400 years
   D. having century years divisible by 400 not be leap years

5. On November 7 sidereal time is about ______ solar time. 5. ___
   A. 3 hours behind B. 3 hours ahead of
   C. 15 hours behind D. 15 hours ahead of

6. In the first part of November an accurate sundial in Philadelphia (longitude - 75°W) reads 12 noon. What is the Central Standard time? 6. ___
   A. 11:43 a.m.
   B. 12:17 p.m.
   C. 10:43 a.m.
   D. 11:17 a.m.

7. What is the longitude if the local mean solar time is 16^h23^m and UT is 5^h00^m? 7. ____
   A. 11^h23^m East B. 11^h23^m West
   C. 21^h23^m East D. 21^h23^m West

8. What is the local hour angle of a star whose right ascension is 2^h30^m if the local sidereal time is 30^m? 8. ____
   A. 2^hEast B. 2^hWest C. 3^hEast D. 3^hWest

9. X-rays have a wavelength that is ________ visible light. 9. ____
   A. longer than
   B. shorter than
   C. equal to
   D. sometimes longer and sometimes shorter than

10. What is the wavelength of radio waves whose frequency is 100 kilocycles, i.e., 10^5c/s? The speed of light is 3×10^{10} cm/sec. 10. ____
   A. $1/3 \times 10^{-5}$ cm B. 3×10^2 cm
   C. 3×10^5 cm D. 3×10^{15} cm

11. What will be the angle that reflected sunlight will make with the surface of a pond if the altitude of the sun is 30°? 11. ____
   A. 0° B. 30° C. 60° D. 90°

12. If the arrows represent the path of a ray of light, then medium A has a ________ index of refraction than medium B. 12. ____

   A. higher
   B. lower
   C. equal
   D. one cannot tell from the information given

13. One loses half of the image if one covers up half of the objective lens of a telescope. 13. ____
   A. True
   B. False
   C. Sometimes true
   D. One cannot tell from the information given

Questions 14-16.
Directions: Consider the images produced by two telescopes for questions 14 through 16.

	A	B
Aperture	10 in.	5 in.
Focal Length	120 in.	40 in.

14. The scale of the images produced by A would be ________ that of B. 14. ____
   A. 1/3 B. 1/2 C. 2 times D. 3 times

15. The image of Mars would be ________ as bright in A as in B. 15. ___
    A. 4/9 B. 9/4 C. 4 times D. 9 times

16. The image of a star would be _____ as bright in A as in B. 16. ___
    A. 4/9 B. 9/4 C. 4 times D. 9 times

17. What magnification would be required to resolve an object 1.2 miles across on the moon with a one-inch telescope on the earth. (The distance of the moon is 240,000 mi.; there are approximately 2 x 10^5 sec of arc in 1 rad.; and assume the unaided eye can resolve objects 1' across). 17. ___
    A. 15X B. 60X C. 2.4 X 10^{12}X
    D. cannot be done no matter how high the magnification

18. Which of the following does one usually NOT have in an equatorial telescope mounting? 18. ___
    A. One axis pointing to the celestial pole
    B. One axis pointing to the zenith
    C. Two mutually perpendicular axes
    D. A clock drive

19. Which type of telescope would be BEST for photographing the spectra of faint stars with a large spectrograph? 19. ___
    A. Refractor B. Reflector C. Schmidt D. Electron

20. What focal length eyepiece should be used to get 100 X magnification with a telescope which has a 5-inch diameter objective with a 50-inch focal length? 20. ___
    A. 1/20 in. B. 1/2 in. C. 2 in. D. 20 in.

21. Which of the following is NOT usually part of a spectrograph? 21. ___
    A. slit B. camera lens C. eyepiece
    D. collimating lens

22. If a star is coming towards us, its spectrum is shifted to 22. ___
    A. shorter wavelengths B. longer wavelengths
    C. ultraviolet D. infrared

23. The reason one cannot photograph extremely fine detail on the moon with the largest telescopes is: 23. ___
    A. Poor optical quality of the telescopes
    B. Insufficient magnification
    C. Turbulence in the earth's atmosphere
    D. Graininess of photographic plates

24. Albedo is 24. ___
    A. the amount of light incident on a body
    B. the fraction of the incident light reflected by a body
    C. the fraction of the incident light absorbed by a body
    D. a measure of the color of a body

25. The moon moves ________ with respect to the stars as seen from the earth. 25. ___

A. $\frac{29\ 1/2}{360}$ °/day
B. $\frac{27\ 1/3}{360}$ °/day
C. $\frac{360}{29\ 1/2}$ °/day
D. $\frac{360}{27\ 1/3}$ °/day

TEST 4

DIRECTIONS: Each question or incomplete statement is followed by several suggested answers or completions. Select the one that BEST answers the question or completes the statement. *PRINT THE LETTER OF THE CORRECT ANSWER IN THE SPACE AT THE RIGHT.*

1. What is the phase of the moon if it is on the meridian at 6 p.m.? 1. ___
   A. new B. 1st quarter C. full D. 3rd quarter

2. If M is the mass of the moon, R its radius, and D its distance from the earth, the surface gravity, g, of the moon is given by 2. ___
   A. $g = \frac{M}{\frac{4}{3}\pi R^3}$
   B. $g = \frac{GM}{R}$
   C. $g = \frac{GM}{D^2}$
   D. $g = \frac{GM}{R^2}$

3. The evidence that the moon does not have an extensive atmosphere consists of all the following EXCEPT 3. ___
   A. occasional mists on floors of some craters
   B. thermal velocity of molecules not much less than the escape velocity
   C. no gradual dimming of stars as they are occulted by the moon
   D. well-defined shadows

4. Suppose a mountain is observed on the moon 100 miles from the terminator. Its shadow is 20 miles long. The diameter of the moon is 2000 miles. How high is the mountain? 4. ___
   A. 1/2 mile B. 1 mile C. 2 miles D. 5 miles

5. The motion of the nodes of the moon's orbit is ______ along the ecliptic. 5. ___
   A. eastward B. westward C. northward D. southward

6. Eclipses of the sun may NOT be used to 6. ___
   A. study the thermal properties of the moon

B. determine the relative locations of the sun, moon, and earth
C. study the corona of the sun
D. study the chromosphere of the sun

7. The eclipse shown is 7. ___
A. an annular eclipse of the sun
B. a partial eclipse of the sun
C. a total eclipse of the sun
D. a partial eclipse of the moon

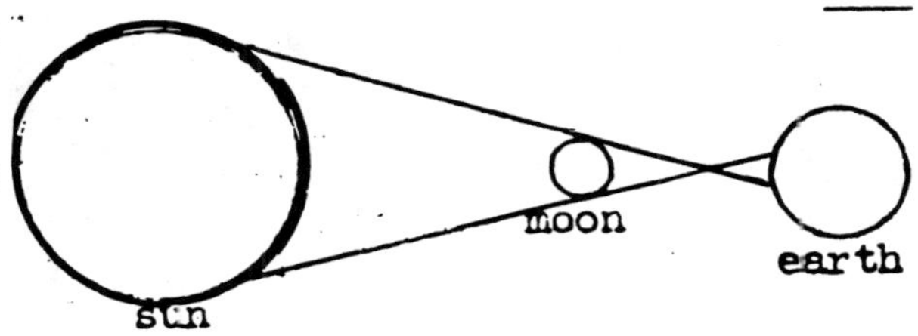

8. An eclipse of the sun can occur only when 8. ___
A. the sun is above the horizon
B. the sun is near the line of nodes of the moon's orbit
C. the moon is full
D. there are no clouds

9. If the semi-major axis of the earth's orbit were 5 AU, but the moon remained at its present distance from the earth, what would the frequency of total eclipses of the sun be? 9. ___
A. more frequent
B. less frequent
C. unchanged
D. one cannot determine from the information given

10. The evidence that tides are caused by the moon and the sun is: 10. ___
A. High tide occurs when the moon is on the meridian
B. Exactly two high tides occur between successive times when the moon is on the meridian
C. The smallest tides occur when the moon is at first and third quarters
D. Choices B and C

11. Tides are produced by 11. ___
A. the moon's pulling the water away from the earth
B. a magnetic attraction of the moon for the water
C. the component of the tidal force perpendicular to the earth's surface
D. the component of the tidal force parallel to the earth's surface

12. An example of a differential gravitational force is 12. ___
A. the difference in force of attraction on one object by two other objects
B. the difference in force of attraction on two objects by a third object
C. a vector

D. the difference between gravitational force and proximity

13. The precession of the earth is caused by 13. ___
    A. the attraction of the sun and moon for the equatorial bulge of the earth
    B. the attraction of the other planets for the earth
    C. the rotation of the earth
    D. all of these

14. The period of precession of the earth is about 14. ___
    A. 1 year B. 50 years C. 360 years D. 26,000 years

15. Which satellite would have the GREATER period, one a million miles from the center of the earth or one a million miles from the center of Jupiter? 15. ___
    A. Earth
    B. Jupiter
    C. both would be equal
    D. one cannot determine from the information given

16. The barycenter of two bodies is 16. ___
    A. the center of one of the bodies
    B. the center of mass (or gravity) of the two
    C. a point halfway between the two
    D. the line joining the two

17. The location of the Trojan Asteroids at two points along Jupiter's orbit is 17. ___
    A. a special case of perturbation theory
    B. a coincidence of orbital distance from the sun, but they could be anywhere along Jupiter's orbit
    C. a special case of the n-body problem
    D. a coincidence of orbital distance from Jupiter, but unrelated to their distance from the sun

18. At what points on the Earth would objects fall straight toward the center of the planet when dropped from a high tower on the surface? 18. ___
    A. At all points B. At the equator only
    C. At the poles only D. At the equator and the poles

19. The tidal bulges are caused by 19. ___
    A. the differential gravitational forces of the sun and the moon
    B. the motions of the sun, Earth and moon producing a centrifugal effect
    C. Both A and B together
    D. Both A and B in combination with the rotation of the Earth

20\. The solid surface of the earth is NOT deformed by 20. ___
A. the rotation of the earth
B. tidal effects due to the moon and the sun
C. the weight of the continental masses
D. the precession of the Earth

21\. The altitude-azimuth coordinate system would be used for 21. ___
A. monitoring the ascent of a weather balloon
B. observing a satellite trajectory
C. determining the relative position of two stars
D. determining the absolute position of a star

22\. The Earth's rotation period with respect to the stars 22. ___
A. was exactly 24 hours in 1900
B. is different at aphelion and perihelion
C. is increasing at 1/1000 second per century
D. is decreased slightly by the precession of the equinoxes

23\. The Earth's period of revolution about the sun with respect to the stars is 23. ___
A. about 365 1/4 solar days
B. the basis of our calendar
C. decreasing at 1/1000 second per century
D. decreased slightly by the precession of the equinoxes

24\. In measuring the distance to the moon by lasar or by radar, we measure the 24. ___
A. speed of the transmitted and reflected pulses
B. intensity of the transmitted and reflected pulses
C. time between the transmitted and reflected pulses
D. wavelength of the transmitted and reflected pulses

25\. In which of the following is there no first, second, third, or last contact? 25. ___
A. Total eclipse
B. Partial eclipse
C. Annular eclipse
D. Penumbral eclipse

TEST 5

Directions: Each question or incomplete statement is followed by several suggested answers or completions. Select the one that BEST answers the question or completes the statement. *PRINT THE LETTER OF THE CORRECT ANSWER IN THE SPACE AT THE RIGHT.*

1\. In order to predict when the moon will first be visible after it is new, one MUST know all of the following EXCEPT the 1. ___
A. apparent angular speeds of the sun and the moon
B. relative distances of the sun and the moon
C. angle between the ecliptic and the horizon
D. angular distance of the moon from the ecliptic

2. The time interval from full moon to full moon is called the ________ period of the moon. 2. ___
   A. sidereal B. synodic C. waxing D. temporal

3. The full moon sets at sunrise. 3. ___
   A. True B. False C. Sometimes
   D. Relative to the time of the month

4. Suppose you observed that a particular star was 20° above the horizon when it was on the meridian. After taking 10^6 steps due north, you observed that this star was only 10° above the horizon when it was on the meridian. What is the circumference of the planet you are on? (Assume the planet is spherical.) 4. ___
   A. 18 X 10^6 steps B. 36 X 10^6 steps
   C. 360 X 10^6 steps D. 7200 X 10^6 steps

5. In order to determine the relative sizes of the sun, moon, and the earth, as was done by Aristarchus, one needs to know all of the following EXCEPT: 5. ___
   A. The relative distances of the sun and the moon from the earth
   B. The angular sizes of the sun and the moon as seen from the earth
   C. The size of the earth's shadow at the distance of the moon, relative to the diameter of the moon
   D. The distance of the moon from the earth

6. The Greeks had to adopt models for the motions of the planets that were more complex than uniform motion in circles centered on the earth because this simple model 6. ___
   A. conflicted with their religious views
   B. did not explain the synodic periods of the planets
   C. did not explain retrograde motion
   D. none of the above

7. Copernicus eliminated all but one of the following features of Ptolemy's model of the planetary system. Which did he NOT eliminate? 7. ___
   A. The equant
   B. The centers of the epicycles of Mercury and Venus being on the Earth-Sun line
   C. The center of **the epicycle-planet line being parallel** to the Earth-Sun line for Mars, Jupiter, and Saturn
   D. Epicycles

8. What is the sidereal period of an inferior planet whose synodic period is 2/3 year? 8. ___
   A. 2/5 year B. 3/5 year C. 5/3 year D. 5/2 year

9. Suppose the greatest elongation of a planet is observed to be 45°, then the planet is ________ planet. 9. ___
   A. An inferior B. A superior C. An intermodal
   D. A double

10. What is the distance of the planet in the previous question from the sun in AU? 10. ___
    A. $1/\sqrt{2}$ AU B. 1 AU C. $\sqrt{2}$ AU D. 2 AU

11. The MAJOR contribution of Tycho Brahe to astronomy was 11. ___
    A. the discovery of the nova in Cassiopea in 1572
    B. the discovery of the great comet of 1577
    C. his observations of the positions of planets
    D. his model of the planetary system

12. What is the eccentricity of the orbit of a planet whose perihelion distance is 1×10^9 miles and whose aphelion distance is 2×10^9 miles? 12. ___
    A. 0 B. 1/3 C. 1/2 D. 3

13. At aphelion the speed of the planet in the previous question will be ______ its speed at perihelion. 13. ___
    A. 1/2 B. the same as C. 2 times D. None of these

14. What is the sidereal period of this planet? (One AU is about 10^8 miles.) 14. ___
    A. $(15)^{2/3}$ year B. $(15)^{3/2}$ year
    C. $(15)^{1/2}$ year D. none of these

15. If you saw an object moving in a circle, you should conclude that 15. ___
    A. a force is acting on the object
    B. the object wants to go in a circle
    C. angels are pushing the object
    D. it is in an equal magnetic field

16. The mass of an object is determined by a spring balance. 16. ___
    A. True B. False C. Possibly
    D. Cannot be determined from the information given

17. In the absence of friction, if you set an object moving on a level surface, it would eventually come to a stop. 17. ___
    A. True B. False C. Possibly
    D. Cannot be determined from the information given

18. It is possible to drive around a banked curve in an icy road without sliding. 18. ___
    A. True B. False C. Possibly
    D. Cannot be determined from the information given

19. Gravitationally, spheres act as point masses. 19. ___
    A. True B. False C. Possibly
    D. Cannot be determined from the information given

20. Gravitation is the strongest force we know. 20. ___
    A. True B. False C. Possibly
    D. Cannot be determined from the information given

21. If one doubles the distance between two masses, the gravitational force between them 21. ____
    A. decreases by a factor of four
    B. decreases by a factor of two
    C. increases by a factor of two
    D. increases by a factor of four

22. If one doubles the mass of one object while keeping the other mass and the distance between the two masses the same, the gravitational force on the first mass 22. ____
    A. decreases by a factor of four
    B. decreases by a factor of two
    C. is unchanged
    D. increases by a factor of two

23. The acceleration experienced by an object in the gravitational field of the earth depends on the mass of the object. 23. ____
    A. True B. False C. Sometimes
    D. Cannot be determined from the information given

24. If the gravitational attraction experienced by one mass as a result of a second mass doubles, then the attraction experienced by the second mass due to the first also doubles. 24. ____
    A. True B. False C. Sometimes
    D. Cannot be determined from the information given

25. Callisto is a satellite of the planet Jupiter. It is about 5 times as far from the center of Jupiter as the moon is from the center of the earth, and its orbital period is about 5/10 that of the moon about the earth. Using this data, what is the ratio of the mass of Jupiter to the mass of the earth? 25. ____
    A. 1/500 B. 1/200 C. 200 D. 500

KEY (CORRECT ANSWERS)

	TEST 1	TEST 2	TEST 3	TEST 4	TEST 5
1.	C	A	D	B	B
2.	A	D	C	D	B
3.	D	B	B	A	A
4.	A	B	C	C	B
5.	B	A	B	B	D
6.	D	C	C	A	C
7.	C	C	A	A	D
8.	D	B	B	B	A
9.	C	A	B	A	A
10.	D	C	C	D	A
11.	C	C	B	D	C
12.	B	A	A	B	B
13.	A	D	B	D	A
14.	D	D	D	D	B
15.	C	A	A	B	A
16.	B	D	C	B	B
17.	B	B	D	C	B
18.	D	C	B	C	A
19.	D	C	B	C	A
20.	B	D	A	D	B
21.	D	C	C	A	A
22.	B	C	A	C	D
23.	A	D	C	A	B
24.	C	D	B	C	A
25.	A	B	D	D	D

ASTRONOMY
EXAMINATION SECTION
TEST 1

Directions: Each question or incomplete statement is followed by several suggested answers or completions. Select the one that BEST answers the question and completes the statement. *PRINT THE LETTER OF THE CORRECT ANSWER IN THE SPACE AT THE RIGHT.*

1. If the speed of an object in a circular orbit around the earth is V mi/hr, then its speed would have to increase to at least ________ in order to escape from the earth. 1. ___
   A. $\sqrt{2}$ V mi/hr B. 2 V mi/hr C. Neither of these
   D. Not enough information

2. In order to determine the age of the earth from the decay of uranium into lead, it is sufficient to know the decay rates of U^{235} and U^{238}, the present concentrations of the isotopes of uranium and lead in the rock sample, and to assume that the lead ore solidified when the earth was formed. 2. ___
   A. True B. False C. Possibly
   D. Cannot be determined from the information given

3. One of the pieces of evidence that the relative positions of the continents may not have been the same in the past as they are at present is the existence of "magnetic anomalies" on the ocean floor. 3. ___
   A. True B. False C. Possibly
   D. Cannot be determined from the information given

4. The interior of the earth is thought to be solid throughout. 4. ___
   A. True B. False C. Possibly
   D. Cannot be determined from the information given

5. One of the proofs that the earth rotates is 5. ___
   A. diurnal motion
   B. the phases of the moon
   C. the aberration of starlight
   D. the behavior of a Foucault pendulum

6. The star Capella (Dec. + 46°) is circumpolar from Ann Arbor,Michigan (Lat.+32°). 6. ___
   A. True B. False C. Possibly
   D. Cannot be determined from the information given

7. The bright star Antares (Dec. -26°) appears to pass overhead at Tucson, Arizona (Lat. + 32°). 7. ___
   A. True B. False C. Possibly
   D. Cannot be determined from the information given

8. One of the proofs that the earth revolves about the sun is 8. ___
   A. diurnal motion
   B. the annual motion of the sun
   C. the aberration of starlight
   D. the behavior of a Foucault pendulum

9. A superior planet has a sidereal period that is ________ an inferior planet. 9. ___
   A. greater than B. less than
   C. equal to D. indefinite when compared to

10. The MAJOR cause of the seasons is the ellipticity of the earth's orbit. 10. ___
   A. True B. False C. Possibly
   D. Cannot be determined from the information given

11. Aurora, "Northern Lights," are caused by 11. ___
   A. thunderstorms
   B. sunlight reflected from the polar icecap
   C. charged particles hitting the earth's atmosphere
   D. vivid imaginations

12. What is the minimum distance from the sun of a planet whose orbit has an eccentricity of 1/4 and a semimajor axis of 4 A.U.? 12. ___
   A. 1 A.U. B. 2 A.U. C. 3 A.U. D. 4 A.U.

13. As seen from the earth, the moon appears to move about ________ with respect to the stars. 13. ___
   A. 1° per day B. 13° per day
   C. 24° per day D. none of these

14. Declination is measured 14. ___
   A. eastward along the celestial equator from the Vernal Equinox
   B. along the horizon from the north point through the east point
   C. along a vertical circle from the horizon
   D. along an hour circle from the celestial equator

15. The presently accepted age of the earth is about 15. ___
   A. 5 million years B. 1 billion years
   C. 5 billion years D. 10 billion years

16. The material in comets is probably MOST like 16. ___
   A. the interstellar medium
   B. the interplanetary medium
   C. present day planetary material
   D. primeval solar system material

17. The material in asteroids is probably MOST like 17. ___
   A. the interstellar medium
   B. the interplanetary medium
   C. present day planetary material
   D. primeval solar system material

18. The following characteristic of the solar system is MAINLY due to the heating effect of the primeval sun. 18. ___
   A. The planets are separated by increasingly large gaps at increasing distance from the sun
   B. Planetary masses are greatest at an intermediate distance from the sun

C. The planets revolve about the sun in the same direction
D. The planets tend to be in nearly circular orbits

19. Jupiter seems to be radiating more heat than it receives from the sun, and appears less frigid than it should be. Without an inner heat source, its temperature depends on 19. ___
A. the amount of solar radiation that strikes it
B. the amount of solar radiation it reflects
C. the amount of solar radiation it absorbs
D. the amount of solar radiation it emits

20. Pluto's orbit is more eccentric and has a higher inclination than any other planet. From this we can deduce that Pluto 20. ___
A. is not an original member of the solar system
B. must have been strongly deflected at some time in order to throw it out of its original circular orbit
C. has not been sufficiently perturbed to bring it into a circular orbit in the ecliptic plane
D. must be a kind of comet

21. Mars has retained much less atmosphere than the earth because its escape velocity 21. ___
A. is less than the molecular speed for most gases, but earth's is greater
B. is greater than the molecular speed for most gases, but earth's is less
C. is less than the molecular speed for most gases, but earth's is much less
D. is greater than the molecular speed for most gases, but earth's is much greater

22. The inside of Saturn's rings rotates faster than the outside. This occurs because 22. ___
A. orbital velocity decreases with increasing distance from the center of mass
B. perturbations are produced by Saturn's moons
C. the equatorial bulge of Saturn produces an additional component of force
D. the atmosphere of Saturn, rotating rapidly with the planet, "drags" the inner part along with it

23. If the spectrum of a planet shows dark lines which are not present in the solar spectrum, 23. ___
A. some gas is present in the sun that is not present in the planet's atmosphere
B. the planet must be rotating
C. some gas is present in the planet's atmosphere that is not present in the sun
D. some gas is present in the planet's atmosphere which is present in the sun

24. The earth's atmosphere is 24. ___
A. uniform in composition at all altitudes
B. non-uniform, having the heaviest molecules at the lowest altitudes
C. non-uniform, having variations in composition due to temperature and radiation

D. non-uniform, having variations in composition due to decreasing escape velocity with altitude

25. The moon's average temperature taken over a long period of time is 25. ___
A. somewhat lower than the earth's average temperature
B. equal to the earth's average temperature
C. somewhat higher than the earth's average temperature
D. undefinable because no atmosphere is present

TEST 2

Directions: Each question or incomplete statement is followed by several suggested answers or completions. Select the one that BEST answers the question or completes the statement. *PRINT THE LETTER OF THE CORRECT ANSWER IN THE SPACE AT THE RIGHT.*

1. The rotation of the earth results in 1. ___
A. a tidal bulge
B. an equatorial bulge
C. a small motion of the north and south poles
D. the shifting of the stars relative to one another

2. The sidereal day is 2. ___
A. always slightly longer than the solar day
B. always slightly shorter than the solar day
C. equal in length to the solar day twice a year
D. always equal to the solar day

3. The "sidereal day" is defined as the interval between two successive transits of ___________ across an observer's meridian. 3. ___
A. a star B. the sun
C. the vernal equinox D. the moon

4. On a day to day basis the passage of time is determined from 4. ___
A. the position of the stars
B. the position of the sun
C. a highly accurate clock
D. a theoretical analysis

5. An eclipse can **NEVER** occur at the 5. ___
A. summer or winter solstice
B. aphelion or perihelion of the earth's orbit
C. time when the moon is near the line of nodes of the earth-moon orbit
D. times of quadrature in the moon's orbit

6. The albedo of the moon is 6. ___
A. the inclination of its orbit
B. the fraction of light it reflects
C. its average temperature
D. the amount of sunlight it receives

7. The path of totality in a solar eclipse is a narrow strip because 7. ___
   A. the moon is much smaller than the sun
   B. the moon is smaller than the earth
   C. the umbra of the moon's shadow barely reaches the earth
   D. the umbra of the moon's shadow is a narrow strip

8. An instantaneous photograph from a weather satellite of a solar eclipse on the earth would show the zone of totality to be 8. ___
   A. the whole earth B. a round spot
   C. a narrow strip D. a circular ring

9. If the orbit of the moon were in the ecliptic plane, we would observe during the course of one month 9. ___
   A. one solar eclipse B. one lunar eclipse
   C. two lunar eclipses D. one solar and one lunar eclipse

10. During the total phase of a solar eclipse, the following CANNOT be seen: 10. ___
    A. Corona B. Planets C. Sunspots D. Stars

11. A solar eclipse can NEVER be seen by 11. ___
    A. an astronaut on the moon
    B. an astronaut in orbit around the earth
    C. both of the above
    D. none of the above

12. Suppose we had "ideal oceans" on the earth's surface. High tides would be experienced when the moon is 12. ___
    A. only at the zenith (point directly overhead)
    B. at the zenith and nadir (point in sky opposite zenith)
    C. on the horizon
    D. none of the above

13. Observations of the tides in the solid earth show it to be 13. ___
    A. viscous, like an oily liquid
    B. slow in adjusting to tidal forces
    C. nearly instantaneous in response to tidal forces
    D. unaffected by tidal forces

14. At any location, the HIGHEST tides occur 14. ___
    A. as a result of lunar action only
    B. as a result of solar action only
    C. at times when the centers of earth, moon, and sun are nearly in line
    D. at times when the moon is directly overhead

15. The earth's non-spherical shape is MAINLY a result of 15. ___
    A. the earth's rotation
    B. tidal distortion

C. perturbations caused by the planets
D. the earth's speed in its orbit

16. The tidal bulge on the earth is always GREATEST 16. ___
A. in the plane of the moon's orbit
B. in the ecliptic plane (plane of the earth's orbit)
C. in the equator plane (plane of the earth's equator)
D. none of the above

17. The length of the day and of the month continually increase due MAINLY to 17. ___
A. perturbations B. aberration
C. tidal friction D. seasonal effects

18. Hipparchus deduced from measurements of seasons 18. ___
A. solar rotation B. precession
C. parallax of stars D. aberration

19. Which of the following has NOT been used to demonstrate rotation of the earth? 19. ___
A. Foucault pendulum B. Foucault gyroscope
C. Foucault knife edge D. Deviation of falling bodies

20. Suppose you were stationed in the crater Copernicus on the moon. The sun would rise 20. ___
A. each earth day B. each earth week
C. each earth month D. never

21. The MAXIMUM number of eclipses (lunar and solar) in a calendar year is 21. ___
A. 2 B. 4 C. 5 D. 7

22. Suppose you were stationed on the crater Copernicus on the moon. The earth would set 22. ___
A. each earth day B. each two weeks
C. each month D. never

23. An observer studying the moon over a period of several years would be able to see 23. ___
A. a quarter of the moon's surface
B. half of the moon's surface
C. more than half of the moon's surface
D. all of the moon's surface

24. When earthshine is brightest on the moon, the phase of the earth as seen from the moon is 24. ___
A. new B. first quarter C. full D. third quarter

25. If the moon revolved from east to west, the synodic month would be ______________ the sidereal month. 25. ___
A. longer than B. shorter than
C. the same as D. dependent upon

TEST 3

Directions: Each question or incomplete statement is followed by several suggested answers or completions. Select the one that BEST answers the question or completes the statement. *PRINT THE LETTER OF THE CORRECT ANSWER IN THE SPACE AT THE RIGHT.*

1. If the earth were to speed up slightly in its orbit, so that a tropical year were completed in exactly 365 days, 3 hours, a leap year would be necessary 1. ___
   A. each four years B. each eight years
   C. each twelve years D. each sixteen years

2. Since the mass of the moon is 1/80th that of the earth, the center of mass of the earth-moon system is 2. ___
   A. nearer to the earth's center than to the moon's center
   B. nearer to the moon's center than to the earth's center
   C. halfway between the two bodies
   D. at the earth's center

3. A body raised high above the earth's surface will weigh 3. ___
   A. less than it would on the surface
   B. more than it would on the surface
   C. the same as it would on the surface
   D. none of the above

4. Stellar parallax, aberration and retrograde motion are the result of the earth's 4. ___
   A. revolution around the sun B. rotation about its axis
   C. precession D. nutation

5. In order to keep the seasons in step with the calendar, one employs the 5. ___
   A. sidereal year B. tropical year
   C. anomalistic year D. martian year

6. If one used the sidereal year as a basis for the calendar, the seasons would fall out of step with the calendar due to 6. ___
   A. the earth's eccentric orbit
   B. the moon's gravitational influence
   C. precession of the equinoxes
   D. none of the above

7. The latitude of the north pole is 7. ___
   A. 0° B. -90° C. +90° D. 360°

8. The sun moves around the sky in a great circle which is called the 8. ___
   A. celestial equator B. zenith
   C. ecliptic pole D. ecliptic

9. The sidereal period of a planet whose distance from the sun is 4 astronomical units is 9. ___
   A. 2 years B. 4 years C. 8 years D. 16 years

10. Over 99% of the mass of the solar system is concentrated in 10. ___
    A. the sun B. the planets
    C. comets and meteors D. none of the above

11. **Aberration** is caused by the 11. ___
    A. inclination of the earth's axis
    B. revolution of the earth around the sun
    C. shape of the earth's orbit
    D. nonspherical shape of the earth

12. The latitude of the equator is 12. ___
    A. 0° B. +90° C. -90° D. 180°

13. Seasons occur because of 13. ___
    A. the rotation of the earth
    B. the changing distance of the earth from the sun
    C. the tilt of the earth's axis
    D. none of the above

14. The GREATEST amount of daylight in Ann Arbor ,**Michigan,** occurs on 14. ___ 21
    A. March 21 B. December 21 C. June 21 D. September 21

15. A total lunar eclipse is visible from 15. ___
    A. the entire earth
    B. a narrow strip on the earth
    C. a small circle on the earth
    D. the entire dark side of the earth

16. A body is dropped down a deep **well** at the north pole. 16. ___
    A. Coriolis effect shifts it to the east.
    B. Coriolis effect shifts it to the west.
    C. Coriolis effect shifts it to the south.
    D. No shift due to Coriolis effect occurs.

17. Variation of latitude causes 17. ___
    A. slight motion of the north celestial pole in the sky
    B. slight error in determination of a ship's position at sea
    C. slight error in the apparent positions of the planets relative to the stars
    D. motion of the air masses over the earth

18. The aberration of starlight results in 18. ___
    A. the apparent motion of stars in "orbits" about nothing
    B. the apparent motion of stars in "orbits" about stars very near to them
    C. an apparent motion of stars about the sun
    D. bending of the light rays from stars

19. Is the sidereal day ever equal to the solar day in the course of a year? 19. ___
    A. Yes B. Sometimes C. No D. During leap years

20. The duration of a second of time is derived from 20. ___
A. an extremely accurate clock
B. the mean position of the sun over a long period of time
C. the mean position of the stars over a long period of time
D. the mean position of the moon over a long period of time

21. We have an extra day in leap year because 21. ___
A. the sidereal year is not equal to the tropical year
B. the tropical year is not an integral number of days
C. the sidereal year is not an integral number of days
D. the tropical year is affected by the precession of the equinoxes

22. Does an eclipse always occur when the moon is exactly on the line of nodes? 22. ___
A. Yes B. No. C. Sometimes
D. During sunspot activity

23. Is there a "Harvest Moon" effect in the spring? 23. ___
A. Yes, in the Northern Hemisphere only
B. Yes, in the Southern Hemisphere only
C. Never
D. Always in both hemispheres

24. Baily's Beads are due to 24. ___
A. irregularities at the sun's edge
B. irregularities at the moon's edge
C. scintillation in the earth's atmosphere
D. the solar corona

25. During totality in a solar eclipse, we cannot see 25. ___
A. prominences on the sun B. the corona of the sun
C. the surface of the sun D. the surface of the moon

TEST 4

Directions: Each question or incomplete statement is followed by several suggested answers or completions. Select the one that BEST answers the question or completes the statement. *PRINT THE LETTER OF THE CORRECT ANSWER IN THE SPACE AT THE RIGHT.*

1. The planets constitute APPROXIMATELY 1. ___
A. the entire mass of the solar system
B. 1/10 of the mass of the solar system
C. 1/100 of the mass of the solar system
D. 1/1000 of the mass of the solar system

2. The mass of Venus is determined from 2. ___
   A. the period and semi-major axis of its orbit around the sun
   B. the observed acceleration it produces on one of its satellites
   C. perturbations in its orbit produced by nearby planets
   D. perturbations in the orbits of small bodies by Venus

3. The rotation of Jupiter is determined by measurement of 3. ___
   A. perturbations in the orbits of its moon
   B. Doppler shifts in the radio emissions of the planet
   C. Doppler shifts in the sunlight reflected from the planet
   D. the precession of its polar axis

4. The key factors which control the nature of a planet's atmosphere are the molecular weights of gases and the planet's 4. ___
   A. surface gravity and mean surface temperature
   B. surfac gravity and albedo
   C. mean surface temperature and albedo
   D. surface gravity, mean surface temperature, and albedo

5. Which of these observed characteristics of Mercury indicates the possible presence of a thin atmosphere? 5. ___
   A. The low albedo of the planet
   B. Absence of a twilight zone
   C. Considerable polarization of the light reflected from the planet
   D. Atmospheric absorption lines are absent from the light reflected from the planet

6. The retrograde rotation of Venus has been accurately measured through the use of radar. The method involves measuring 6. ___
   A. radar radiation originating at fixed points on the planet's surface
   B. the effect radar signals from earth have on the planet's atmosphere
   C. radar signals from earth reflected by fixed features on the planet's surface
   D. radar radiation originating in the sun reflected by fixed features on the planet's surface

7. The transition from winter to summer on Mars is marked by 7. ___
   A. decrease of the southern polar cap and darkening of the maria
   B. decrease of the southern polar cap and lightening of the maria
   C. increase of the southern polar cap and darkening of the maria
   D. increase of the southern polar cap and lightening of the maria

8. The "red spot" on Jupiter is 8. ___
   A. a transient cloud of vast size photographed in 1931
   B. a permanent fixed feature of the planet
   C. a permanent but variable feature of the planet
   D. a wandering feature of constant size and color

9. Saturn's rings are not a solid sheet of material. Which of the following is NOT a proof of this? 9. ___
   A. When viewed edge on, they cannot be discerned.
   B. Stars can be seen through the rings.
   C. Doppler shifts in the spectrum of light reflected by the rings show the inside rotating faster than the outside.
   D. Cassini's division occurs at a distance where particles would be perturbed severely by the satellite Mimas.

10. Uranus differs markedly from the other "gas giants" in 10. ___
   A. its albedo B. its period of rotation
   C. the inclination of its axis
   D. the eccentricity of its orbit

11. The gas giants -- Jupiter, Saturn, Uranus, and Neptune -- are considerably 11. ___
   A. more massive and denser B. more massive but less dense
   C. less massive but denser D. less massive and less dense

12. Venus and earth are a pair of planets similar to each other except for 12. ___
   A. period of rotation B. diameter
   C. mass D. inclination of axes

13. Jupiter and Saturn are a pair of planets similar to each other EXCEPT for 13. ___
   A. period of rotation B. diameter
   C. mass D. color

14. Uranus and Neptune are a pair of planets similar to each other. They differ significantly from Jupiter and Saturn in diameter, mass, *and* 14. ___
   A. period of rotation B. color
   C. eccentricity of orbit D. internal structure

15. Which of the following is NOT true of Neptune? 15. ___
   A. One of its moons moves backwards in its orbit.
   B. One of its moons has the highest eccentricity of any satellite
   C. It does not fit in the planetary positions predicted by Bode's law.
   D. It is the most oblate planet.

16. Pluto's rotation period is obtained by 16. ___
   A. observation of surface features over a long period
   B. measurement of Doppler shift in the sunlight reflected by the planet

C. observation of variations of the planet's brightness
D. fluctuations in radio emission from the planet

17. Which of the following is NOT true of the minor planets? 17. ___
A. They generally have highly eccentric orbits.
B. The largest is almost 500 miles in diameter.
C. Estimates of their numbers exceed 40,000.
D. Most of them have orbits of low inclination.

18. The mass of a minor planet is obtained from 18. ___
A. perturbations in its orbit produced by large planets
B. the period and semi-major axis of its orbit
C. its estimated size and assumed density
D. its period of rotation and estimated size

19. Bode's law indicates a planetary position that is occupied by some of the minor planets. The estimated combined mass of all the minor planets is 19. ___
A. almost that of Jupiter
B. about 1/3 that of the earth
C. about 1/20 that of the moon
D. about 1/1600 that of the moon

20. Which of the planets have two known natural satellites? 20. ___
A. Mars and Uranus B. Mars and Neptune
C. Mars, Uranus, and Neptune D. Mars alone

21. The D, E, F1, and F2 layers in the earth's atmosphere are concentrations of 21. ___
A. ozone molecules
B. hydrogen atoms
C. ionized oxygen and nitrogen atoms
D. charged dust particles

22. Which of the following is NOT a proof of the lack of a thin lunar atmosphere? 22. ___
A. Absence of a twilight zone
B. Presence of numerous craters
C. Sudden blinking out of a star during occultation by the moon
D. Sudden drop in temperature of most of the surface during an eclipse of the moon

23. The large lunar maria are 23. ___
A. younger than the adjacent craters
B. the same age as the adjacent craters
C. older than the adjacent craters
D. of indeterminant relative age

24. Craters on the moon are generally believed to be 24. ___
A. volcanic in origin
B. the result of meteoritic impact explosions
C. formed by geological forces when the crust formed

D. any of the above, depending on individual circumstances

25. The earth's magnetic field is probably due to 25. ___
 A. large amounts of magnetized iron in the semi-molten core
 B. rapid rotation of magnetized iron in the semi-molten mantle
 C. electrical currents generated by rotation in the semi-molten mantle
 D. electrical currents generated by rotation in the semi-molten core

TEST 5

Directions: Each question or incomplete statement is followed by several suggested answers or completions. Select the one that BEST answers the question or completes the statement. *PRINT THE LETTER OF THE CORRECT ANSWER IN THE SPACE AT THE RIGHT.*

1. The EARLIEST existing astronomical records are 1. ___
 A. Chinese B. Egyptian C. Greek D. Arabic

2. The sun moves with respect to the stars 2. ___
 A. west-east only
 B. north-south only
 C. west-east rapidly, north-south slowly
 D. erratically

3. Greek astronomy was practiced CHIEFLY in 3. ___
 A. Greece B. Egypt C. Arabia D. Mesopotamia

4. The sun appears to rise in the _______ part of the sky. 4. ___
 A. eastern B. southern C. western D. northern

5. How many moving astronomical objects can be seen regularly without a telescope? 5. ___
 A. 2 B. 4 C. 7 D. 11

6. Constellations are 6. ___
 A. clusters of stars B. bright stars
 C. groups of planets D. areas of the sky

7. The apparent form of the moon as seen in the sky depends on 7. ___
 A. how much of the moon's daylight hemisphere is visible
 B. the rate of rotation of the moon
 C. the rate of revolution of the moon
 D. the motion of the earth

8. At first-quarter the moon has the shape of 8. ___
   A. a quarter-circle B. a half-circle
   C. a crescent D. a rectangle

9. The words new, crescent, first quarter, full, gibbous, third quarter, ALL describe lunar 9. ___
   A. mountains B. eclipses C. phases D. tides

10. An observation, known to Aristotle, in support of the spherical shape of the earth, is the 10. ___
    A. shape of shadow of earth on moon during a lunar eclipse
    B. phases of the moon
    C. rotation of the earth
    D. alternation of seasons

11. Aristarchus' principal astronomical measurement was 11. ___
    A. relative size of system earth-moon-sun
    B. stellar parallax
    C. rotation of earth
    D. obliquity of ecliptic

12. Eratosthenes' principal astronomical measurement was 12. ___
    A. size of moon B. planetary distances
    C. size of earth D. planetary periods

13. Planets generally move with respect to the stellar background 13. ___
    A. not at all B. from west to east
    C. from east to west D. in undetermined fashion

14. Temporary planetary motion from east to west for short intervals of time is called 14. ___
    A. precession B. rotation
    C. dispersion D. retrograde

15. What percentage fraction of the total area of the sky is visible to an observer located on earth's equator? 15. ___
    A. 10% B. 25% C. 50% D. 100%

16. Which of the following systems is geocentric? 16. ___
    A. Solar system B. Jupiter's satellite system
    C. Saturn's satellite system D. Earth-moon system

17. Copernicus based his theories on observations by 17. ___
    A. himself
    B. Greek astronomers
    C. Arabian astronomers
    D. combinations of Greek and Arabian astronomers

18. A planet that is farther from the sun than the earth is called a ___________ planet. 18. ___
    A. superior B. inferior C. minor D. Copernican

19. A superior planet may be observed BEST at 19. ___

A. inferior conjunction B. quadrature
C. noon D. opposition

20. An inferior planet may be observed BEST at 20. ___

A. inferior conjunction B. greatest elongation
C. noon D. superior conjunction

21. One pair of the following named planets are inferior planets. Which pair? 21. ___

A. Saturn, Jupiter B. Mercury, Venus
C. Mars, Venus D. Saturn, Mercury

22. The period of revolution of a planet around the sun from a star to the same star again, as seen from the sun, is the 22. ___

A. synodic period B. sidereal period
C. harmonic period D. natural period

23. The interval between two successive times when a planet occupies the same position in relation to the sun as seen from the earth is called the 23. ___

A. synodic period B. sidereal period
C. harmonic period D. natural period

24. To derive the relative distances of the planets, which of the following quantities need NOT be known? 24. ___

A. Sidereal period
B. Synodic period
C. Angles of elongation at two times
D. Brightness

25. Tycho Brahe gained his principal reputation as 25. ___

A. an observing astronomer B. a mathematician
C. a physicist D. a cosmogonist

KEY (CORRECT ANSWERS)

TEST 1		TEST 2		TEST 3		TEST 4		TEST 5	
1.	A	1.	B	1.	B	1.	D	1.	A
2.	B	2.	B	2.	A	2.	D	2.	C
3.	A	3.	C	3.	A	3.	C	3.	B
4.	B	4.	C	4.	A	4.	A	4.	A
5.	D	5.	D	5.	B	5.	C	5.	C
6.	B	6.	B	6.	C	6.	C	6.	D
7.	B	7.	C	7.	C	7.	A	7.	A
8.	C	8.	B	8.	D	8.	C	8.	B
9.	A	9.	D	9.	C	9.	A	9.	C
10.	B	10.	C	10.	A	10.	C	10.	A
11.	C	11.	D	11.	B	11.	B	11.	A
12.	C	12.	B	12.	A	12.	A	12.	C
13.	B	13.	C	13.	C	13.	C	13.	B
14.	D	14.	C	14.	C	14.	B	14.	D
15.	C	15.	A	15.	D	15.	D	15.	D
16.	D	16.	A	16.	D	16.	C	16.	D
17.	C	17.	C	17.	B	17.	A	17.	D
18.	B	18.	B	18.	A	18.	C	18.	A
19.	C	19.	C	19.	C	19.	C	19.	D
20.	C	20.	C	20.	B	20.	B	20.	B
21.	D	21.	D	21.	B	21.	C	21.	B
22.	A	22.	D	22.	B	22.	B	22.	B
23.	C	23.	C	23.	B	23.	A	23.	A
24.	C	24.	C	24.	B	24.	B	24.	D
25.	A	25.	B	25.	C	25.	D	25.	A

ASTRONOMY
EXAMINATION SECTION
TEST 1

Directions: Each question or incomplete statement is followed by several suggested answers or completions. Select the one that BEST answers the question or completes the statement. *PRINT THE LETTER OF THE CORRECT ANSWER IN THE SPACE AT THE RIGHT.*

1. Kepler's GREATEST contributions to astronomy are 1. ___
   A. music of the spheres
   B. optical theory
   C. tables of planets
   D. three laws of planetary motion

2. Which of the following is NOT a property of an ellipse? 2. ___
   A. Plane curve
   B. Has two foci
   C. sum of the distances from the two foci to any point on the curves is constant
   D. The center of the ellipse is the same distance from all points of the curve

3. Kepler's FIRST LAW was deduced from observation of 3. ___
   A. all planets B. Mars only
   C. the inferior planets D. Sun and Mars

4. Kepler's SECOND LAW is based on the following: 4. ___
   A. Mars moves fastest when closest to the sun
   B. Properties of the ellipse
   C. Retrograde motions
   D. Mars' period of revolution

5. Kepler's THIRD LAW resulted from a search for 5. ___
   A. underlying harmony in nature
   B. gravitational forces
   C. a method of measuring planetary distances
   D. a method of improving planetary periods

6. Kepler revised the estimated distance to the sun. His revision was based on 6. ___
   A. his three laws B. direct measurement
   C. no direct evidence D. no diurnal parallax for Mars

7. One of the following was an astronomical discovery NOT made by Galileo with a telescope. Indicate the correct one. 7. ___
   A. Phases of Venus B. Spots on the sun
   C. Mountains on the moon D. Faculae on the sun

8. Rockets and jets operate on the principle of 8. ___
   A. Newton's FIRST LAW B. Newton's SECOND LAW
   C. Newton's THIRD LAW D. Universal Law of Gravitation

9. On the surface of the earth, Neil Armstrong weighs 180 pounds, whereas on the surface of the moon he weighs 30 pounds. The reason for this is: 9. ___
   A. He is less massive on the moon.
   B. The moon is more massive than the earth.
   C. The moon has no atmosphere.
   D. The moon is less massive than the earth.

10. When dropped from the same height, a cannon ball and a marble strike the earth's surface at the same time. This is TRUE because 10. ___
   A. the mass of the marble and the cannon ball are the same
   B. the weight of the marble and the cannon ball are the same
   C. the acceleration of the marble and the cannon ball are the same
   D. the gravitational force on the marble and the cannon ball are the same

11. The number of meteoritic impact craters known on the earth is 11. ___
   A. two
   B. several dozen
   C. several hundred
   D. several thousand

12. Mercury rotates in such a way that the same side faces the sun 12. ___
   A. at all times
   B. at successive perihelions
   C. at alternate perihelions
   D. on every third perihelion

13. Which of the following is NOT a proof that Venus has a significant atmosphere? 13. ___
   A. Its measured temperature is different at different wavelengths.
   B. It reflects 75% of the sunlight striking it.
   C. The points of its crescent extend more than half way around the planetary disc.
   D. The spectrum of light reflected from it has lines in it not in the solar spectrum.

14. The "canals" on Mars seen by observers on earth have 14. ___
   A. been observed by the Mariner IV space probe
   B. not been observed by Mariner IV, and are thought to be an optical illusion
   C. not been observed by Mariner IV because it did not look at regions supposedly having "canals"
   D. not been observed by Mariner IV because it was specially equipped with filters to show craters, not "canals"

15. Which of the following is NOT true of Jupiter? 15. ___
   A. It has two satellites, Ganymede and Callisto, larger than the earth's moon.
   B. It has a magnetic field and "Van Allen" belts.
   C. Its atmosphere consits mainly of hydrogen, helium, carbon dioxide, methane, and ammonia.
   D. Its atmosphere is probably not more than several hundred miles thick.

16. Saturn's rings MOST probably formed 16. ___
   A. when the planet itself formed
   B. over billions of years as the planet "captured" such material
   C. when a comet was perturbed into that orbit around the planet
   D. when a moon moved closer to the planet than Roche's limit

17. Which of the following is NOT true of Pluto? 17. ___
    A. Its orbit has the highest inclination of any planet.
    B. Its orbit has the greatest eccentricity of any planet.
    C. It is smaller than Mars in diameter and mass.
    D. One of its moons moves backwards in its orbit.

18. A meteor shower radiant, seen against the celestial sphere, is 18. ___
    A. the angle from the zenith of the convergent point of the meteor tracks
    B. the small region from which the meteor tracks radiate
    C. the great circle to which all meteor tracks are parallel
    D. the center of the large circle which defines the region of the meteor shower

19. The ages of meteorites, measured by radioactive dating, are 19. ___
    A. considerably younger than the age of the earth
    B. about the same as the age of the earth
    C. considerably older than the age of the earth
    D. various, some younger and some older

20. Which of the following is NOT true of comets? 20. ___
    A. They probably have a rocky or metallic core.
    B. The greater part of their substance is in the form of ices which sublime as they approach the sun.
    C. Cometary masses are known from the perturbation they produce on asteroids and satellites of planets.
    D. Comets have been measured near the sun with temperatures as high as 4500°K.

21. The age of the earth is derived from studies of the radioactivity of surface rocks. A rock's radioactive "life" as measured this way begins 21. ___
    A. in the primordial "fireball" -- the beginning of the universe
    B. when the solar system formed from dust and gas
    C. when the rock solidified from the liquid magma which formed the primitive planet (earth)
    D. when the rock was last exposed to strong atomic radiation from the new formed sun

22. Which of the following is NOT primarily due to atomic particles in the earth's magnetic field? 22. ___
    A. The aurora borealis (Northern lights)
    B. The solar wind
    C. Static in ordinary radio transmission
    D. The Van Allen belts

23. Primary evidence that the moon's core has solidified is the 23. ___
    A. absence of lunar seismic activity
    B. lower average temperature of the surface compared to the earth

C. lack of a lunar magnetic field
D. large variation of temperature on the surface

24. There are fewer large craters in the moon's maria because 24. ___
A. the maria remained molten much longer than other areas after the moon formed
B. the maria consist of a material that does not form craters easily
C. the maria are of more recent origin and insufficient time has passed for many large object collisions
D. the maria are of more recent origin and large object collisions in the solar system no longer occur

25. The light colored streaks radiating out from the crater Copernicus indicate that 25. ___
A. the material underneath the surface was lighter colored than the surface
B. the material underneath the surface was darker colored than the surface
C. the impact explosion produced a chemical change in the ejected material which lightened its color
D. blast burning of surface material occurred in the impact explosion, producing the streaks

TEST 2

Directions: Each question or incomplete statement is followed by several suggested answers or completions. Select the one that BEST answers the question or completes the statement. *PRINT THE LETTER OF THE CORRECT ANSWER IN THE SPACE AT THE RIGHT.*

1. Perturbations of the orbit of Uranus lead to the discovery of 1. ___
A. Halley's comet B. Neptune
C. Titan D. Jupiter

2. Consider a minor planet in a circular orbit about 4 times as far from the sun as is the earth. The orbital period of this minor planet is 2. ___
A. 4 years B. 8 years C. 16 years D. 64 years

3. Suppose the earth were twice as far from the sun as it is now. The sun's attractive force on the earth would then be ______ as great as now. 3. ___
A. 1/2 B. 1/4 C. 1/8 D. 1/16

4. One way to reduce environmental pollution on the earth would be to send all industrial wastes into the sun. In order to send a spaceship directly to the sun, it MUST be launched 4. ___
A. from the equator at local noon on March 21 or September 22

B. in the direction of the earth's motion at the escape velocity of about 7 miles/second
C. opposite to the direction of the earth's orbital motion at the escape velocity of 7 miles/second
D. opposite to the direction of the earth's orbital motion at the earth's orbital velocity of about 18.5 miles/second

5. The mass of the earth can be measured by 5. ___
A. comparing the attraction between the earth and a given mass, M_A, to the attraction between another mass, M_B, and M_A
B. multiplying the density of a small sample of the surface of the earth by the volume of the earth
C. measuring the altitude of stars from different, known positions on the earth
D. none of the above

6. How long **will it** take a space probe to go from the Earth to Jupiter on a minimum energy transfer orbit? The semi-major axis of Jupiter's orbit is about 5 AU, and of the Earth's orbit is 1 AU. 6. ___
A. $\frac{1}{2}(4^{3/2})$ year.   B. $3^{3/2}$ year
C. $\frac{1}{2}(3^{3/2})$ year   D. $\frac{1}{2}(6^{3/2})$ year

7. How much would a 100-pound person weigh on Jupiter? The mass of Jupiter is about 300 times that of the Earth, and its radius is about 11 times that of the Earth. G is the gravitational constant. 7. ___
A. $100 \times \frac{300}{(11)}$ lb.   B. $100 \times \frac{300}{(11)^2}$ lb.
C. $100 \times \frac{G \times 300}{(11)^2}$ lb.   D. $100 \times \frac{G \times 300 \times 100}{(11)^2}$ lb.

8. One of the proofs that the earth revolves about the sun is 8. ___
A. diurnal motion
B. the annual motion of the sun
C. the aberration of starlight
D. the behavior of a Foucault pendulum

9. Why isn't Venus always exactly behind the sun at superior conjunction? 9. ___
A. Its period is too long.
B. It is always exactly behind the sun at superior conjunction.
C. It is never behind the sun at superior conjunction.
D. Its orbital plane is inclined to the ecliptic.

10. As seen from the earth, the moon appears to move about ________ with respect to the stars. 10. ___

A. 1° per day　　B. 13° per day
C. 24° per day　　D. none of the above

11. If the speed of an object in a circular orbit around the earth is V mi/hr, then its speed would have to increase to at least _________ in order to escape from the earth. 11. ___
   A. $\sqrt{2}$ V mi/hr　　B. 2 V mi/hr
   C. Neither of these　　D. Not enough information

12. Ariel is a satellite of the planet Uranus. It is about 1/2 as far from the center of Uranus as the Moon is from the center of the Earth, and its orbital period is about 1/10 that of the Moon about the Earth. Using this data, what is the ratio of the mass of Uranus plus Ariel to the mass of the Earth plus the Moon? 12. ___
   A. 4/1000　　B. 8/100　　C. 100/8　　D. 1000/4

13. The Greeks had to adopt models for the motions of the planets that were more complex than uniform motion in circles centered on the earth because this simple model 13. ___
   A. conflicted with their religious views
   B. did not explain the synodic periods of the planets
   C. did not explain retrograde motion
   D. none of the above

14. A body falling freely in a gravitational field has 14. ___
   A. constant velocity
   B. constant acceleration
   C. neither of the above
   D. both of the above

15. What is the eccentricity of the orbit of a planet whose perihelion distance is 1 AU and whose aphelion distance is 3 AU? 15. ___
   A. 1/4　　B. 3/2　　C. 1/2　　D. 4

16. As seen from the north ecliptic pole, which of the following statements is TRUE? 16. ___
   A. The orbital motions of the inferior planets are counter-clockwise and of the superior planets are clockwise.
   B. The orbital motions of the superior planets are counter-clockwise and of the inferior planets are clockwise.
   C. The orbital motions of all the planets are counter-clockwise.
   D. The orbital motions of all the planets are clockwise.

17. Suppose Aristarchus had found that the interval from the first quarter moon to third quarter moon was the same as from the third quarter to the first. He would have concluded that 17. ___
   A. the sun and moon were equally distant from the earth

B. the sun was infinitely far away
C. the moon was infinitely far away
D. the sun was $\sqrt{2}$ times as far from the earth as the moon

18. Copernicus determined the distances of inferior planets from the sun (in terms of the earth's distance from the sun) by observing the planet at 18. ___
A. quadrature
B. inferior conjunction
C. superior conjunction
D. greatest elongation

19. If at sunrise the moon were setting for an observer in northern Michigan, the phase of the moon would be 19. ___
A. new B. first quarter C. full D. last quarter

20. Suppose you observed that a particular star was 20° above the horizon when it was on the meridian. After taking 10^6 steps due north, you observed that this star was only 10° above the horizon when it was on the meridian. What is the circumference of the planet you are on? (Assume the planet is spherical.) 20. ___
A. 18×10^6 steps
B. 36×10^6 steps
C. 369×10^6 steps
D. 7200×10^6 steps

21. The polar axis of a telescope 21. ___
A. always points toward the observer's zenith
B. is always set parallel to the axis of rotation of the earth
C. is oriented the same, relative to the ground, at all observatories
D. is parallel to the telescope's other axis, the declination axis

22. When moonshine is brightest on the earth, what must be the phase of the earth, as seen from the moon? 22. ___
A. New B. First quarter C. Full D. Last quarter

23. The solar spectrum observed at the earth is 23. ___
A. a bright line (emission) spectrum superimposed on a continuous spectrum
B. a dark line (absorption) spectrum superimposed on a continuous spectrum
C. a continuous spectrum only
D. both A and B

24. The accepted age of Stonehenge was recently increased by about 800 years. This change came about because it is thought now that the carbon-14 method of dating organic material gives results that are
A. in error due to variations in the rate of production of carbon 14 in the past
B. in error due to variations in the decay rate of carbon 14 from sample to sample

C. in error due to inaccurate measuring techniques
D. carbon-14 has nothing to do with it; the Druids were discovered to have lived earlier than was previously thought

25. The solar day (24 hours of solar time) is slightly shorter than the sidereal day (24 hours of sidereal time). 25. ___
A. Always B. Never C. Sometimes D. Relative

TEST 3

Directions: Each question or incomplete statement is followed by several suggested answers or completions. Select the one that BEST answers the question or completes the statement. *PRINT THE LETTER OF THE CORRECT ANSWER IN THE SPACE AT THE RIGHT.*

1. Apparent solar days are NOT of uniform duration throughout the year because 1. ___
   A. the earth's rotation rate is speeding up
   B. the earth's axis is not perpendicular to the plane of the earth's orbit
   C. the earth's orbital speed is faster at perihelion than at aphelion
   D. both B and C

2. Suppose you were in a ship sailing east and you cross the International Date Line. Before you cross it, the standard time (zone time) is 9:45 Monday morning. What are the standard time and day 15 minutes later after you cross the line? 2. ___
   A. 10 a.m. Sunday B. 11 a.m. Sunday
   C. 9 a.m. Tuesday D. 10 a.m. Tuesday

3. Assuming Newton's laws of mechanics are correct, the Foucault pendulum experiment 3. ___
   A. is a definite proof that the earth rotates
   B. is fun to look at but does not tell us anything
   C. does not show the earth's rotation at the pole
   D. works well to demonstrate the earth's rotation at the equator

4. Right ascension and declination, which are based on an extension of the earth's axis of rotation and equatorial plane onto the celestial sphere, form a practical coordinate system because 4. ___
   A. the definitions are easy to remember
   B. the coordinates of a fixed point on the celestial sphere, e.g., a star, are not changed by the earth's rotation
   C. the coordinates of a fixed point on the celestial sphere change constantly due to the earth's rotation

D. Polaris is near the pole and another bright star is near the Vernal Equinox

5. The equation of time 5. ___
   A. is difficult to solve since it is an n-th order polynomial with constant coefficients
   B. is Newton's form of Kepler's second law (equal areas)
   C. is a relation that gives the obliquity of the ecliptic of any date
   D. is the amount by which the fictitious (mean) and true (apparent) sun differ in their positions in the sky

6. A blackbody has a ________ type of spectrum. 6. ___
   A. line emission B. band emission
   C. continuous emission D. line absorption

7. Telescope A: A reflector with a 30-inch aperture objective of 150-inch focal length and an eyepiece of 2-inch focal length 7. ___

   Telescope B: A refractor with a 10-inch aperture objective of 150-inch focal length and an eyepiece of 1-inch focal length

   Which statement below is FALSE?
   A. A has better resolving power than B.
   B. B has greater magnification than A.
   C. Point sources will be brighter in B.
   D. Extended sources will be brighter in A.

8. How can the mid-summer sunrise be directly over the heel-stone at Stonehenge now even though the stone has tilted since the monument was built 40 centuries ago? 8. ___
   A. Planetary precession has caused the apparent position of sunrise to shift enough to compensate for the change in position of the top of the stone.
   B. Luni-solar precession has caused the time of the mid-summer sunrise to change enough to compensate for the change in position of the top of the stone.
   C. The Druids are the only ones to observe the sunrise, and their reports are not to be trusted.
   D. The mid-summer sunrise did not take place over the heelstone when the monument was built.

9. Atmospheric refraction makes a star's altitude appear less than it really is. 9. ___
   A. True B. False C. Sometimes D. Relative

10. Which of the following is MOST closely related to the fact that the moon appears dimly red during many total lunar eclipses? 10. ___
   A. "Earthshine"

B. Diffraction of light near the earth's edge
C. Refraction of light by the earth's atmosphere
D. Infrared radiation from the moon

11. If the temperature of a blackbody were doubled, the rate at which it radiates energy would increase by a factor of ______________. 11. ___
A. $2^{2/3}$ B. 2 C. $2^{3/2}$ D. 2^4

12. If the earth and the moon were farther from the sun (but the earth-moon distance remained constant), then 12. ___
A. solar eclipses would occur more often
B. solar eclipses would occur less often
C. solar eclipses would be visible over a larger area of the earth
D. both A and D

13. Would a company whose goal was to make a perpetual motion machine be a good investment? 13. ___
A. Yes, since everyone likes to get something for nothing, such a machine would sell well.
B. Yes, such a machine would end worries about running out of energy sources on the earth.
C. No, according to our experience, summarized in the second law of thermodynamics, it is not possible to build such a machine.
D. No, such a machine would not be beneficial because it would be too hard to control; there would be no way to stop it.

14. The harvest moon is caused by 14. ___
A. the inclination of the celestial equator to the horizon
B. the inclination of the moon's orbit to the ecliptic
C. the autumnal equinox
D. the clear skies in the fall

15. You are aboard a ship which is hijacked by pirates who leave you stranded on a desert island. You must send a message for help by bottle, but you do not know where you are. You have a watch that gives you Greenwich time and you see that at night the north star is about 1/4 of the way between the horizon and the zenith (but closer to the horizon than the zenith). Can you estimate your location? 15. ___
A. No, not enough information
B. Yes, 5^h east longitude, 23° north latitude
C. Yes, 5^h east longitude, 23° south latitude
D. Yes, 5^h west longitude, 23° north latitude

16. Spectra of Jupiter show strong absorption due to ammonia. This absorption is not present in spectra of Saturn. It is thought that this is due to 16. ___

A. Saturn's being so cold that the ammonia would be frozen on the surface of the planet
B. Saturn's not having any ammonia
C. Saturn's rotating so fast that the ammonia lines are so broad that they cannot be recognized
D. the ammonia lines of Saturn being hidden by absorption due to ammonia in the earth's atmosphere

17. A similarity between the rings of Saturn and the minor planet zone is 17. ___
A. their location
B. the orbital periods of the particles in them
C. the size distribution of the particles in them
D. the existence of gaps or regions where there are very few particles

18. Long-period comets generally have their orbital planes nearly coincident with the ecliptic plane. 18. ___
A. True B. False C. Sometimes
D. Not enough information given

19. Short-period comets were 19. ___
A. long-period comets whose orbits decayed by interaction with the solar wind
B. long-period comets that were perturbed by stars into short-period orbits
C. long-period comets that were perturbed by the planets into short-period orbits
D. formed in their present, short-period orbits

20. The accretion theory for the origin of the comets implies that 20. ___
A. comets are formed in long-period orbits of high eccentricity
B. comets are formed in long-period orbits of low eccentricity
C. comets are formed of material left over from the formation of the solar system
D. we will eventually run out of comets

21. The period of rotation of Mercury is now thought to be 2/3 of its orbital period. This recent value for the period of rotation is due to 21. ___
A. extensive visual observations
B. extensive radar observations
C. information returned by mariner spacecraft
D. none of the above

22. The atmosphere of Venus appears to be MOSTLY 22. ___
A. water vapor B. oxygen
C. nitrogen D. carbon dioxide

23. Not much is known about the surface of Venus because the planet 23. ___
    A. is rotating too fast to photograph the surface
    B. is too close to the sun
    C. has a cloudy atmosphere
    D. is too far away to resolve the surface features

24. Meteorites are all so small that they are not capable of damaging areas on the earth as large as a city block. 24. ___
    A. True B. False C. Not until the present time
    D. Unknown

25. The ________ of at least some meteorites indicate that they were once part of planet-sized objects. 25. ___
    A. sizes B. masses C. internal structures
    D. orbits

TEST 4

Directions: Each question or incomplete statement is followed by several suggested answers or completions. Select the one that BEST answers the question or completes the statement. *PRINT THE LETTER OF THE CORRECT ANSWER IN THE SPACE AT THE RIGHT.*

1. The temperature of a planet is determined by the ________ between the energy received from the sun (minus that reflected) and the energy radiated by the planet. 1. ___
    A. balance B. interference
    C. compensation D. none of these

2. The Roche limit is 2. ___
    A. the limiting brightness which the eye can detect
    B. the limiting distance inside which large solid bodies would be destroyed by tidal forces
    C. the zone inside which rapidly moving charged particles are trapped in the earth's magnetic field
    D. the limit at which solar energy vaporizes

3. The gegenshine is
    A. a meteor shower
    B. a flow of light opposite the sun
    C. the upper part (above 50 miles) of the earth's atmosphere
    D. interior crust of the lunar surface

4. Upper limits have been placed on the masses of comets using 3. ___
    A. perturbation theory
    B. estimates of comets' dimensions
    C. Kepler's Third Law
    D. Galileo's First Hypothesis

5. One of the major consequences of the theory that the planetary system arose as a natural result of the formation of the sun is that 5. ___
   A. it removes the angular momentum difficulty
   B. it shows that Kepler's laws are valid elsewhere
   C. it implies that there are probably planets around other stars
   D. it proves the theories of the evolution of life

6. It is thought that life could have originated spontaneously on the primitive earth provided there was (were) 6. ___
   A. a reducing (hydrogen rich) atmosphere
   B. water
   C. energy in the form of lightning or ultraviolet light
   D. all of these

7. The MAIN reason for believing there is probably not intelligent life on Mars is that 7. ___
   A. we have not seen it
   B. it has not contacted us
   C. there is too little water on Mars
   D. it is too cold on Mars

8. One objection to the idea that the moon was captured by the earth is that the temperatures of the surfaces of the earth and the moon would have been raised to the melting point of rock, but no evidence has been found that this occurred. This objection has recently been removed 8. ___
   A. by noting that a resonance between the rotation of the earth and the revolution of the moon would prevent such strong heating
   B. by noting that the age of the moon is much less than formerly thought
   C. by interpreting the peculiar glazing phenomenon found by the Apollo 11 astronauts on lunar rocks as evidence that such strong heating did occur
   D. by discovering that the moon really came from the crust of the earth from what is now the Pacific Ocean

9. Which of the following CAN be used to determine the relative ages of lunar features? 9. ___
   A. Overlap of features
   B. Number of craters per unit area
   C. Sharpness of craters
   D. All of these

10. Is it likely that there are bright aurora on Mars as there are on the Earth? 10. ___
    A. Yes B. No C. Maybe D. Cannot be determined

11. It now seems that the canals on Mars are 11. ___
    A. irrigation ditches

B. more or less linear arrangements of craters
C. dark dust blown by the prevailing winds
D. inexplicable

12. Of the following, the STRONGEST evidence that the south pole cap of Mars is frozen carbon dioxide is 12. ___
A. the temperature of the cap is that expected for frozen carbon dioxide
B. the color is that of frozen carbon dioxide
C. no rivers of water are seen leading away from the cap
D. it is probably not frozen carbon dioxide

13. The rotation rate of Jupiter has been determined MOST accurately by 13. ___
A. radar
B. doppler shift in visible spectra
C. timing the passage of features in the cloud belts across the central meridian of the planet
D. measuring the oblateness of the planet

14. The evidence that Jupiter and Saturn are MOSTLY hydrogen is 14. ___
A. their large size
B. their large mass
C. their light color
D. their low density

15. The MOST recently discovered satellite of Saturn, its tenth, was predicted by 15. ___
A. studying the structure of the ring system
B. applying Bode's Law to Saturn's satellites
C. noting that 10 is an even number but 9 is an odd number
D. Albert Einstein

16. The fact that Venus, Mars and the Moon have vanishingly small magnetic fields, while the Earth and Jupiter have strong magnetic fields implies that 16. ___
A. magnetic fields are a property of all rotating bodies
B. in order to have a strong magnetic field, a planet must rotate rapidly
C. in order to have a strong magnetic field, a planet must have a mass of the same order as that of the earth or larger
D. both B and C

17. The oblateness of Jupiter and Saturn is caused by 17. ___
A. their distance from the sun
B. their extensive system of satellites
C. their thick cloud layers
D. their rapid rotation

18. MOST of the minor planets have orbits whose semi-major axes are about 18. ___
A. 1AU B. 1.5 AU C. 3 AU D. 5 AU

19. The Trojans are NOT 19. ___
    A. between the orbits of Jupiter and Mars
    B. 60° from Jupiter as seen from the sun
    C. an example of a special solution of the three-body problem
    D. minor planets

20. Comets always have tails. 20. ___
    A. True B. False C. True so far D. Unknown

21. A comet tail that is relatively straight and blue is probably a tail of 21. ___
    A. type I (ionized gases) B. type II (dust)
    C. type III (water) D. none of the above

22. A meteor shower results from 22. ___
    A. a statistical fluctuation (concentration) in the random arrival of interplanetary particles
    B. observing in the hours after midnight
    C. the earth's passing through the orbit of a comet
    D. passing through a geophysical phenomenon

23. The meteors after midnight are more spectacular than before midnight because 23. ___
    A. the particles collide head-on with the earth
    B. the orbital speed of the earth is greater
    C. the particles are bigger
    D. the particles catch up with the earth

24. The day on Mars is ___________ as the day on earth. 24. ___
    A. less than half as long
    B. about the same length
    C. more than twice as long
    D. three times as long

25. The planets (including the minor planets) are the only objects in the solar system whose spacing is given approximately by "Bode's Law" or by a rule which is similar to Bode's Law. 25. ___
    A. True B. False C. True so far D. Unknown

KEYS (CORRECT ANSWERS)

	TEST 1	TEST 2	TEST 3	TEST 4
1.	D	B	D	A
2.	D	B	D	B
3.	D	B	A	B
4.	A	D	B	A
5.	A	A	D	C
6.	D	C	C	D
7.	D	B	C	C
8.	C	C	A	A
9.	D	D	B	D
10.	C	B	C	B
11.	B	A	D	B
12.	C	C	D	A
13.	A	C	C	C
14.	B	B	B	D
15.	C	C	A	A
16.	D	C	A	D
17.	D	B	D	D
18.	B	D	B	C
19.	B	C	C	A
20.	C	B	A	B
21.	C	B	B	A
22.	B	A	D	C
23.	A	B	C	A
24.	C	A	B	B
25.	A	B	C	B

ASTRONOMY
EXAMINATION SECTION
TEST 1

Directions: Each question consists of a statement. You are to indicate whether the statement is TRUE (T) or FALSE (F). *PRINT THE LETTER OF THE CORRECT ANSWER IN THE SPACE AT THE RIGHT.*

1. The earliest existing astronomical records known at the time of writing Abell's text were of Chinese origin. 1. ___
2. The clear weather in the region of the Sahara desert played a more important part in early astronomy than it does in modern astronomy. 2. ___
3. The sun's principal apparent movement with respect to the stars is from west to east. 3. ___
4. In winter the sun rises nearer to the north point of the horizon than in summer. 4. ___
5. The apparent shape of the illuminated part of the moon as seen in the sky depends upon the difference in the direction from earth to sun and the direction from earth to moon. 5. ___
6. At first quarter, the moon has the shape of a quarter circle. 6. ___
7. Crescent, first quarter, full, gibbous, third quarter are words describing eclipses. 7. ___
8. Phases of the moon are evidence of the spherical shape of the earth. 8. ___
9. A computation of the size of the earth was a result of Eratosthenes's observation of the sun. 9. ___
10. Planets generally move from east to west with respect to the stellar background. 10. ___
11. A retrograding planet moves from east to west with respect to the stellar background. 11. ___
12. The solar photosphere is hotter than the sunspot umbra. 12. ___
13. If the largest possible fraction of the sky is to be observed, an observatory should be near the equator. 13. ___
14. Copernicus's theories were based on his own astronomical observations. 14. ___
15. "Superior" and "inferior" refer to distances of planets from the sun, with respect to the earth. 15. ___
16. Inferior planets are most easily observed in conjunction with the sun. 16. ___
17. Superior planets are most easily observed at opposition. 17. ___
18. Pluto is called an inferior planet because it is one of the smallest planets. 18. ___
19. The sideral period of revolution of a planet is the time 19. ___

required for the planet to move from conjunction with a star to conjunction with the same star as viewed from the sun.

20. All of the following planets, Mercury, Venus, Mars, Jupiter, Saturn, Uranus, Neptune and Pluto may appear in conjunction with the sun. 20. ___

21. Full moon occurs at intervals determined by the moon's synodic period of revolution about the earth. 21. ___

22. The synodic period of Mars is assumed to be exactly two years. The sidereal period then is two years. 22. ___

23. The period of a planet whose distance from the sun is four astronomical units is eight years. Assume a circular orbit. 23. ___

24. Telescopes were made by opticians for other uses before they were used in astronomy. 24. ___

25. Kepler's GREATEST contributions to astronomy were based on Brahe's precision observations. 25. ___

TEST 2

Directions: Each question consists of a statement. You are to indicate whether the statement is TRUE (T) or FALSE (F). *PRINT THE LETTER OF THE CORRECT ANSWER IN THE SPACE AT THE RIGHT.*

1. One of the modern forms of telescope was invented by Kepler. 1. ___

2. Objects weigh less on the moon than on earth because the moon is less dense than earth. 2. ___

3. When dropped from the same height, a cannon ball and a buck-shot strike earth's surface at the same time. The explanation is the same gravitational acceleration of the buck-shot and the cannon ball. 3. ___

4. Rotation of the earth causes points on the earth's equator to be more distant from the earth's center than a point at the north or south pole. 4. ___

5. Aberration of starlight is a small apparent shift in the position of a star in a direction opposite to the direction of the earth's motion. 5. ___

6. An eclipse CANNOT occur at winter solstice. 6. ___

7. The albedo of the moon is its average temperature. 7. ___

8. The path of totality of a solar eclipse is a narrow strip in the position of a star in a direction opposite to the direction of earth's motion. 8. ___

9. An instantaneous photograph taken from an artificial satellite of a total solar eclipse on earth would show the zone of totality as a nearly circular spot. 9. ___

10. During the total phase of a solar eclipse, sunspots can be observed within the zone of totality. 10. ___

11. The sidereal day is always slightly longer than the solar day. 11. ___

12. The sidereal day is defined as the interval between two successive transits of the vernal equinox. 12. ___

13. Assume the plane of the moon's orbit to be in the plane of the ecliptic. During one sidereal month there would be two lunar eclipses. 13. ___

14. Suppose the earth were completely covered with an ideal ocean. High tide at a given location would occur when the moon is on the horizon. 14. ___

15. Observations of tides in solid earth show that the earth is as rigid as steel. 15. ___

16. The earth's shape is MAINLY determined by its own gravitational attraction and its rotation. 16. ___

17. The barycenter of the earth-moon system is halfway to the moon from the earth. 17. ___

18. The length of a day is very gradually increasing. The increase is caused MAINLY by attractions of the other planets on the earth. 18. ___

19. Measurements of the duration and the beginnings of the seasons led to the discovery of precession. 19. ___

20. The rotation of a Foucault pendulum is a proof of the rotation of the earth. 20. ___

21. As viewed from a crater near the moon's equator, the earth rises above the lunar horizon each earth month. 21. ___

22. There may be as many as seven lunar and solar eclipses in one calendar year. 22. ___

23. An observer mapping the moon from an observatory on 23. ___

earth would be able to see in several years'time more than half of the lunar surface.

24. Earth-shine is brightest on the moon when the earth is at quarter phase as seen from the moon. 24. ___

25. If the moon revolved in its orbit in a direction opposite to its actual revolution, the synodic month would be longer than the sidereal month. 25. ___

TEST 3

Directions: Each question consists of a statement. You are to indicate whether the statement is TRUE (T) or FALSE (F). *PRINT THE LETTER OF THE CORRECT ANSWER IN THE SPACE AT THE RIGHT.*

1. If one revolution of the earth around the sun took exactly 365 days, 3.000 hours, a leap year would be necessary each eight years. 1. ___

2. A mass raised high above the earth's surface will weigh more than on earth. 2. ___

3. Stellar parallax, retrograde motions, and aberration are effects of earth's revolution around the sun. 3. ___

4. The latitude of the north pole is 0°. 4. ___

5. The ecliptic plane is the plane containing the earth's orbit. 5. ___

6. The planets constitute APPROXIMATELY 10% of the entire mass of the solar system. 6. ___

7. Sunspots are randomly distributed over the solar disk. 7. ___

8. The retrograde rotation of Venus has been measured by radar signals from the earth that are reflected by fixed features on Venus's surface. 8. ___

9. Some seasonal changes on Mars are marked by a shrinking of the south polar cap and a darkening of the maria. 9. ___

10. The great red spot on Jupiter is a permanent feature of the planet which varies in size and color. 10. ___

11. Saturn's rings are not solid sheets of material since stars are visible when occulted by the rings. 11. ___

12. The inner planets, Mercury, Venus, Earth, Mars, are considerably less massive but denser than the outer planets. 12. ___

13. Venus and Earth are nearly alike in mass and diameter but greatly different in rate of rotation. 13. ___

14. Uranus and Neptune are nearly alike. They are similar to Jupiter and Saturn in mass. 14. ___

15. The period of rotation of the planet Pluto has been obtained from observations of periodic variations of the brightness of the planet. 15. ___

16. The LARGEST of the asteroids, Ceres, has a diameter of almost 500 miles. 16. ___

17. The mass of an asteroid is estimated from its size and density. 17. ___

18. Bode's law has a planetary position near the average of the positions of many asteroids. 18. ___

19. The estimated combined masses of all the asteroids is about that of the moon. 19. ___

20. The radio-reflecting layers of earth's atmosphere are concentrations of ionized atoms. 20. ___

21. Many more impact craters would be present on the moon if it had a dense atmosphere. 21. ___

22. Meteoritic impacts and some volcanic action seem to have formed the craters on the moon. 22. ___

23. There is no direct evidence of **collisions of meteors** with the earth. 23. ___

24. The sun is a huge "nuclear reactor" whose major energy source is the fusion of hydrogen to form helium. 24. ___

25. "Canals" on Mars have not been observed from space vehicles. 25. ___

TEST 4

Directions: Each question consists of a statement. You are to indicate whether the statement is TRUE (T) or FALSE (F). *PRINT THE LETTER OF THE CORRECT ANSWER IN THE SPACE AT THE RIGHT.*

1. Jupiter has a magnetic field and the associated "Van Allen" belts. 1. ___

2. Pluto is at present believed to be comparable with Mars in diameter. 2. ___

3. The radiant of a meteor swarm is a small area of the sky from which the meteors apparently diverge. 3. ___

4. The ages of meteors are comparable with the age of the earth 4. ___

5. Comets are sources of all meteorites. 5. ___

6. If the moon is observed to rise at midnight, its phase is full. 6. ___

7. The moon is the largest satellite in the solar system, in relation to its primary, earth. 7. ___

8. The Jovian planets are composed of heavier elements than those that compose the earth. 8. ___

9. The photosphere,chromosphere,ionosphere, and corona are all parts of the solar atmosphere. 9. ___

10. Only the very largest meteors can reach the surface of the earth. 10. ___

11. The earth reflects more sunlight than the moon. 11. ___

12. A magnetic compass could be used on the moon to indicate directions from crater to crater. 12. ___

13. The MOST abundant constituent of the earth's atmosphere is water vapor. 13. ___

14. The core of the earth is the same density as the surface rocks. 14. ___

15. The core of the earth is hotter than the surface rocks. 15. ___

16. Slipping of rocks in fault zones causes earthquakes. 16. ___

17. The mantle is the outermost layer of earth. 17. ___

18. The age of the earth is believed to be about five billion years. 18. ___

19. Lunar maria as viewed from the earth are smoother than other parts of the moon's surface. 19. ___

20. Lunar maria are lighter in color than most other parts of the moon's surface. 20. ___

21. Aurorae are caused by charged particles encountering the earth's atmosphere. 21. ___

22. The lunar maria are generally younger than the lunar mountains. 22. ___

23. Lunar rays are MOST easily seen during full moon. 23. ___

24. The moon rotates once in the same time it takes to revolve once about the earth. 24. ___

25. Mercury rotates on its axis with the same period as its period of revolution around the sun. 25. ___

TEST 5

Directions: Each question consists of a statement. You are to indicate whether the statement is TRUE (T) or FALSE (F). *PRINT THE LETTER OF THE CORRECT ANSWER IN THE SPACE AT THE RIGHT.*

1. The surface of Mercury is a good conductor of heat. 1. ___

2. The mass and temperature of Mercury make it possible for this planet to hold an atmosphere. 2. ___

3. Venus always appears as the morning star. 3. ___

4. The surface of Venus is cooler than the surface of the earth. 4. ___

5. Two satellites have been discovered revolving around Venus. 5. ___

6. The MAJOR component of Venus's atmosphere is nitrogen. 6. ___

7. The atmosphere of Venus is denser than earth's. 7. ___

8. Artificial satellites near Venus have detected a magnetic field on the planet. 8. ___

9. Mars has two satellites. 9. ___

10. Mars is not an identical twin of the earth, it is smaller. 10. ___

11. Comets always have tails. 11. ___

12. The Martian atmosphere is denser than earth's atmosphere. 12. ___

13. Mars's atmosphere is composed MAINLY of carbon dioxide. 13. ___

14. Large yellow clouds on Mars are in composition much like clouds of water vapor on earth. 14. ___

15. Jupiter has one dozen known satellites. 15. ___

16. Jupiter's diameter is one hundred times earth's diameter. 16. ___

17. Jupiter's mean density is less than earth's mean density. 17. ___

18. The solar chromosphere is hotter than the corona. 18. ___

19. The white bands on Jupiter are MOSTLY ammonia particles. 19. ___

20. Radio observations show that Jupiter has a magnetic field. 20. ___

21. If it were possible to find an ocean large enough to hold it, the planet Saturn would float. 21. ___

22. Cassini's division is a gap between the two outermost rings of Saturn. 22. ___

23. Saturn's atmosphere is very much like the atmosphere of Jupiter. 23. ___

24. If Venus were at western elongation, it would be a "morning star." 24. ___

25. One of the pieces of evidence that the relative positions of the continents may not have been the same in the past as they are at present is the existence of "magnetic anomalies" on the ocean floor. 25. ___

TEST 6

Directions: Each question or incomplete statement is followed by several suggested answers or completions. Select the one that BEST answers the question or completes the statement. *PRINT THE LETTER OF THE CORRECT ANSWER IN THE SPACE AT THE RIGHT.*

1. Suppose you were an astronaut in orbit about the earth and you wish to overtake another spacecraft in the same orbit as you but some distance ahead. Assuming you are facing the direction you are traveling, your FIRST maneuver should be 1. ___
   A. to fire rockets with their exhaust directed behind you
   B. to fire rockets with their exhaust directed in front of you
   C. to fire rockets with their exhaust directed below you
   D. none of the above

2. What is the eccentricity of the orbit of a planet whose maximum and minimum distance from the sun are 1 AU and 1/2 AU? 2. ___
   A. 1/3 B. 1/2 C. 2 D. 3

3. The MOST important contribution of Copernicus to astronomy was to 3. ___
   A. discover the relation between sidereal and synodic period
   B. eliminate epicycles
   C. put the sun, not the earth, as the central object in the solar system
   D. determine the relative distances of the planets

4. Which of the following was NOT areason advanced by Aristotle to prove that the earth was round? 4. ___
   A. The earth's shadow on the moon during a lunar eclipse was always round
   B. If one went far enough in one direction, one came back to where he started
   C. Different stars were visible in Greece and in Egypt
   D. Elephants had been observed in both India and Morocco

5. Hipparchus discovered the precession of the equinoxes by 5. ___
   A. observing the motions of the stars
   B. comparing his coordinates for stars with those of earlier observers
   C. observing the motions of the sun
   D. adding epicycles to the model of Eudoxus

6. An example of a superior planet is 6. ___
   A. Mars B. Venus C. Earth D. Moon

7. Suppose an inferior planet has a synodic period of two years. What would its sidereal period be? 7. ___
   A. 1/2 year B. 2/3 year C. 3/2 year D. 2 years

8. The effect of having uniform motion take place about a point that was displaced from the central object in the models of the "solar system" due to both Ptolemy and Copernicus was to approximate the motion summarized in Kepler's __________ law. 8. ___
   A. first B. second C. third D. none of the above

9. As seen from the earth, direct motion of a planet is 9. ___
   A. from west to east B. from east to west
   C. from north to south D. from south to north

10. Can the synodic period of a planet be less than 1 year? 10. ___
   A. Yes B. No C. Sometimes
   D. It cannot be determined from the information given

11. What is the mass of a celestial object that has a satellite whose period of revolution is 10 times that of the moon around the earth and whose orbit is 100 times larger than that of the moon? 11. ___
    A. 10^{-4} X mass of the earth
    B. 10^{-1} X mass of the earth
    C. 10 X mass of the earth
    D. 10^{4} X mass of the earth

12. The evidence that the earth has a magnetic field is 12. ___
    A. magnetic compasses work B. the Van Allen layers
    C. aurora D. all of these

13. The altitude of the north celestial pole is equal to 13. ___
    A. the latitude of the observer B. 0° C. 90° D. 42°

14. A proof that the earth revolves around the sun is 14. ___
    A. the change of direction of a Foucault pendulum
    B. the seasons
    C. the phases of Venus
    D. aberration of starlight

15. What is the azimuth of a star directly in the east and halfway from the horizon to the zenith? 15. ___
    A. 45° B. 90° C. 270° D. 0°

16. What is the right ascension of the vernal equinox? 16. ___
    A. 0^h B. 6^h C. 12^h D. 18^h

17. If the year were 720 days long, how far would the sun move each day with respect to the stars (as seen from the earth)? 17. ___
    A. 1/2° B. 1° C. 2° D. none of these

18. In their studies of the heavens, the Greeks tried to 18. ___
    A. predict the occurrence of "guest stars"
    B. make geometrical models of celestial objects
    C. predict the future positions of celestial objects by arithmetic techniques
    D. make geometrical models of the motions of celestial objects

19. You are on a strange planet. You note that the stars do not rise or set, but circle around parallel to the horizon. Then you travel over the surface of the planet in a straight line for 1,000 miles and find that at this new place the stars rise straight up from the horizon in the east and set straight down in the west. What is the circumference of the planet? 19. ___
    A. 1,200/2π miles B. 2π X 1,000 miles
    C. 4,000 miles D. 4,000/2π miles

20. About what time of the day or night does the moon rise when it is at first quarter? 20. ___
A. Midnight B. Sunrise C. Noon D. Sunset

21. Suppose you were in a plane flying east and you cross the International Date Line. Before you cross it, the standard time is 9:45 Monday morning. What are the standard time and day 15 minutes later after you cross the line? 21. ___
A. 10 a.m. Sunday B. 11 a.m. Sunday
C. 9 a.m. Tuesday D. 10 a.m. Tuesday

22. In what month would a star whose right ascention is 12^h be on the meridian at midnight local mean time? 22. ___
A. March B. September
C. neither of these D. both A and B

23. The sidereal day is ____________ the solar day. 23. ___
A. shorter than B. longer than
C. equal to D. dependent upon

24. At the present time the MOST accurate way to measure the passage of time is 24. ___
A. with a quartz crystal clock
B. by the rotation of the earth
C. by occultations of stars by the moon
D. with an atomic clock

25. Suppose a planet where 2 AU from the sun. It would receive ________ as much energy per unit area from the sun as does the earth. 25. ___
A. 1/4 B. 1/2 C. 2 times D. 4 times

KEYS (CORRECT ANSWERS)

TEST 1		TEST 2		TEST 3		TEST 4		TEST 5		TEST 6	
1.	TT	1.	F	1.	T	1.	T	1.	F	1.	B
2.	TT	2.	F	2.	F	2.	T	2.	F	2.	A
3.	TT	3.	T	3.	T	3.	T	3.	F	3.	C
4.	FF	4.	T	4.	F	4.	T	4.	F	4.	B
5.	T	5.	F	5.	T	5.	F	5.	F	5.	B
6.	F	6.	F	6.	F	6.	F	6.	F	6.	A
7.	F	7.	F	7.	F	7.	T	7.	T	7.	B
8.	F	8.	T	8.	T	8.	F	8.	F	8.	B
9.	T	9.	T	9.	T	9.	F	9.	T	9.	B
10.	F	10.	F	10.	T	10.	F	10.	T	10.	A
11.	T	11.	F	11.	T	11.	T	11.	F	11.	D
12.	T	12.	T	12.	T	12.	F	12.	F	12.	D
13.	T	13.	F	13.	T	13.	F	13.	T	13.	A
14.	F	14.	F	14.	F	14.	F	14.	F	14.	C
15.	T	15.	T	15.	T	15.	T	15.	T	15.	B
16.	F	16.	T	16.	T	16.	T	16.	F	16.	A
17.	T	17.	F	17.	T	17.	F	17.	T	17.	A
18.	F	18.	F	18.	T	18.	T	18.	F	18.	D
19.	T	19.	T	19.	F	19.	T	19.	T	19.	C
20.	T	20.	T	20.	T	20.	F	20.	T	20.	C
21.	T	21.	F	21.	F	21.	T	21.	T	21.	A
22.	T	22.	T	22.	T	22.	T	22.	T	22.	A
23.	T	23.	T	23.	F	23.	T	23.	T	23.	A
24.	T	24.	F	24.	T	24.	T	24.	T	24.	D
25.	T	25.	F	25.	T	25.	F	25.	T	25.	A

EXAMINATION SECTION

TEST 1

DIRECTIONS: Each question or incomplete statement is followed by several suggested answers or completions. Select the one that BEST answers the question or completes the statement. *PRINT THE LETTER OF THE CORRECT ANSWER IN THE SPACE AT THE RIGHT.*

1. We see one-half of the lighted portion of the moon at the 1.____
   A. first quarter B. full moon
   C. new crescent D. new moon

2. A planet whose orbit is between the sun and the orbit of the earth is 2.____
   A. Jupiter B. Mars C. Pluto D. Venus

3. The changing position of the stars during the night is MAINLY the result of the 3.____
   A. *inclination* of the earth's axis
   B. *rotation* of the earth
   C. *rotation* of the stars
   D. *revolution* of the earth

4. The gravitational force of the moon is exerted 4.____
   A. only upon the side of the earth nearest the moon
   B. only upon the point on the earth nearest the moon
   C. upon the center of the earth only
   D. upon the entire earth

5. The work done by tides is gradually slowing down the earth's period of rotation. 5.____
   It is, therefore, reasonable to predict that millions of years from now,
   A. the earth will be farther away from the moon than now
   B. the frequency of tides will be greater than now
   C. days will be shorter than they are now
   D. phases of the moon will change more rapidly than now

6. The earth's gravitational attraction is GREATEST at 6.____
   A. the North Pole
   B. the Equator
   C. a point on the earth's surface directly under the moon
   D. a point on the earth's surface exactly opposite the moon

7. The twinkling of a star is caused by 7.____
   A. the star itself
   B. interplanetary dust
   C. defects in the structure of the human eye
   D. turbulence within the atmosphere

8. The small dipper seems to turn about the North Star once each day because 8.____
   A. all stars move in great circles on the celestial sphere
   B. the earth turns on its axis
   C. the North Star is the last star in the handle of the dipper
   D. the planets revolve around the sun

9. The stars that comprise the Big Dipper are of which one of the following magnitudes? 9.____
   A. +1 B. +2 C. +3 D. +4

10. The distance of any remote group of stars can be determined if the group contains which one of the following? 10.____
   A. Double stars B. Dwarf stars
   C. Super giant stars D. Cepheid variable stars

11. Among the following, the expression that BEST describes the typical star in old age is 11.____
   A. blue giant B. blue-white intermediary
   C. yellow giant D. red dwarf

12. According to Hubble, the galaxies are BEST classified into which one of the following sets of categories? 12.____
   A. Elliptical and spherical
   B. Elliptical and spiral
   C. Spherical and spiral
   D. Elliptical, spherical, and spiral

13. Any given point on the moon receives continuous light for a number of days CLOSEST to which one of the following? 13.____
   A. 7 B. 10 C. 14 D. 29

14. If the first quarter moon rises at noon, the last quarter moon will rise at a time CLOSEST to which one of the following? 14.____
   A. Midnight B. 3 A.M. C. 3 P.M. D. 6 P.M.

15. The concept that distant stars are receding from the earth at great velocity is implied from observation of the 15.____
   A. Cepheid variables
   B. parallax
   C. red giants changing toward white
   D. spectroscopic shift to the red

16. On the eastern coast, the altitude of the North Star is CLOSEST to which one of the following? 16.____
   A. 39° B. 42° C. 45° D. 47½°

17. Which one of the following constellations seems to dip below the horizon in New York City? 17.____
   A. Cassiopeia B. Cepheus C. Orion D. Ursa Major

18. When Venus is in line with the earth and the sun, on the side of the sun nearest the earth, it is 18.____
   A. at inferior conjunction B. at superior conjunction
   C. in opposition D. at an elongation of 180°

19. Studies of Cepheid variable stars have been of MOST value in revealing which one of the following? 19.____
   A. Nature of atomic fusion B. Magnitude of stars
   C. Distance of galaxies D. Size of galaxies

20. Which one of the following statements is TRUE regarding the tide-producing force of the moon? It varies 20.____
   A. *directly* with the distance
   B. *directly* with the square of the distance
   C. *inversely* with the square of the distance
   D. *inversely* with the cube of the distance

21. Which one of the following constellations is CLOSEST to the celestial north pole? 21.____
   A. Ursa Minor B. Ursa Major C. Cepheus D. Draco

22. Which one of the following discovered the harmonic law relating distance of a planet to its time of revolution? 22.____
   A. Tycho Brahe B. Kepler C. Galileo D. Newton

23. As nearly as can be estimated, the center line of the Milky Way is the 23.____
   A. celestial equator B. plane of the ecliptic
   C. celestial meridian D. galactic equator

24. A planet which is NEVER visible at night is 24.____
   A. Mars B. Neptune C. Mercury D. Saturn

25. The planetoids are located between 25.____
   A. Earth and Mars B. Mars and Venus
   C. Mars and Jupiter D. Jupiter and Saturn

KEY (CORRECT ANSWERS)

1. A
2. D
3. B
4. D
5. A

6. A
7. D
8. B
9. B
10. D

11. D
12. B
13. C
14. A
15. D

16. B
17. C
18. A
19. C
20. D

21. A
22. B
23. D
24. C
25. C

TEST 2

DIRECTIONS: Each question or incomplete statement is followed by several suggested answers or completions. Select the one that BEST answers the question or completes the statement. *PRINT THE LETTER OF THE CORRECT ANSWER IN THE SPACE AT THE RIGHT.*

1. At which one of these phases does the moon rise APPROXIMATELY at midnight and set APPROXIMATELY at noon? 1.___
   A. New
   B. Full
   C. First quarter
   D. Last quarter

2. The planetoids are located with GREATEST frequency between 2.___
   A. Earth and Mars
   B. Jupiter and Saturn
   C. Mars and Venus
   D. Mars and Jupiter

3. When the moon is between the earth and the sun and in line with them, the tidal range on the earth at any one place is 3.___
   A. average
   B. less than average
   C. more than average
   D. at maximum

4. All stars visible to the naked eye belong to 4.___
   A. the Milky Way galaxy
   B. a number of galaxies
   C. the solar system
   D. the Milky Way galaxy and several spiral nebulae

5. Of the following planets, the one that has its orbital plane MOST inclined to the ecliptic is 5.___
   A. Mercury
   B. Earth
   C. Saturn
   D. Pluto

6. Among the characteristics listed, the one in which stars differ LEAST is 6.___
   A. diameter
   B. mass
   C. density
   D. distance from the earth

7. The NEAREST star to the earth is 7.___
   A. the sun
   B. the moon
   C. Polaris
   D. Alpha Centauri

8. Among the following, the statement about TIROS that is NOT true is that 8.___
   A. the name is an acronym for Television and Infra-Red Observation Satellite
   B. it is used principally for weather observation
   C. it is powered by nuclear batteries
   D. it is unmanned

9. Primary cosmic rays are believed to consist MOSTLY of which one of the following? 9.___
   A. Photons
   B. Protons
   C. Mesons
   D. Neutrons

10. Among the following, the planet that has the MOST eccentric orbit is 10.____
 A. Mercury B. Earth B. Saturn D. Pluto

11. Of the following, the color exhibited by the HOTTEST star is 11.____
 A. red B. yellow C. blue-white D. orange

12. Spring tides are MOST likely to occur at the phase of the moon called 12.____
 A. first quarter B. full moon
 C. gibbous D. crescent

13. If the moon were to stop rotating, we would 13.____
 A. see both sides of it
 B. have higher tides
 C. have full moon continuously
 D. have half moon continuously

14. The North Star is found in which one of the following constellations? 14.____
 A. The Little Dipper B. The Big Dipper
 C. Orion D. Arcturus

15. At the autumnal equinox, the vertical rays of the sun at noon are over the 15.____
 A. Tropic of Cancer B. Tropic of Capricorn
 C. Arctic Circle D. Equator

16. The planet whose diameter is CLOSEST in size to that of the earth is 16.____
 A. Mercury B. Venus C. Mars D. Neptune

17. The time required for the sun to pass a given meridian twice in succession is called a ____ day. 17.____
 A. mean solar B. sidereal C. solar D. civil

18. The altitude of the North Star is CLOSEST to 65° at which one of the following locations? 18.____
 A. Arctic Circle B. City of New York
 C. North Pole D. Tropic of Cancer

19. If a full moon occurs on June 1st, the next new moon phase will take place at a time CLOSEST to which one of the following? 19.____
 A. June 8 B. June 15 C. June 22 D. June 29

20. Circular star trails are evidence that the 20.____
 A. earth is revolving B. stars are revolving
 C. stars are rotating D. earth is rotating

21. Of the following heavenly bodies, the one which has come CLOSEST to the earth in historic times is a 21.____
 A. moon B. meteor C. planet D. comet

22. The Milky Way is classified as which one of the following? 22.___
A. Star B. Solar system C. Galaxy D. Constellation

23. Of the following, the BEST description of the earth's orbit is that it is 23.___
A. circular B. highly elliptical
C. slightly elliptical D. three-dimensional

24. The moon's orbit around the earth is BEST described as a(n) 24.___
A. ellipse B. straight line
C. hyperbola D. circle

25. When the moon rises approximately at noon and sets at about midnight, its phase is 25.___
A. new moon B. last quarter
C. first quarter D. full moon

KEY (CORRECT ANSWERS)

1. D
2. D
3. D
4. A
5. D

6. B
7. A
8. C
9. B
10. D

11. C
12. B
13. A
14. A
15. D

16. B
17. C
18. A
19. B
20. D

21. B
22. C
23. C
24. A
25. C

TEST 3

DIRECTIONS: Each question or incomplete statement is followed by several suggested answers or completions. Select the one that BEST answers the question or completes the statement. *PRINT THE LETTER OF THE CORRECT ANSWER IN THE SPACE AT THE RIGHT.*

1. A planet which is NOT visible to the naked eye is 1.____
   A. Saturn B. Neptune C. Mars D. Venus

2. All of the stars visible to the naked eye in our sky are members of 2.____
   A. the solar system B. a group of galaxies
   C. the Milky Way galaxy D. the Andromeda galaxy

3. Spring tides occur when 3.____
   A. the gravitational forces of the sun and moon affect the earth in the same direction
   B. the sun's gravitational force acts at right angles to that of the moon
   C. the moon is in the new crescent phase
   D. none of the above

4. Of the following stars, the one FARTHEST from the earth is 4.____
   A. Sirius B. Polaris C. Proxima Centauri D. Betelgeuse

5. The FAINTEST star that the unaided eye can see is classified as a star of the ____ magnitude. 5.____
   A. first B. second C. third D. sixth

6. The GREATEST source of radio energy yet discovered by radio astronomy lies in the constellation 6.____
   A. Orion B. Ursa Major C. Cyngus D. Canis Major

7. The new-moon phase occurs when the moon is at 7.____
   A. conjunction B. opposition
   C. elongation D. quadrature

8. The BRIGHTEST of all the *fixed stars* observed from earth is 8.____
   A. Betelgeuse B. Formalhaut C. Sirius D. Vega

9. An eclipse of the moon occurs when the 9.____
   A. moon passes between the earth and the sun
   B. sun passes between the earth and the moon
   C. earth passes between the sun and the moon
   D. new moon phase occurs

10. A reading taken in New York City would indicate that the magnetic north pole is located ____ the geographic north pole. 10.____
    A. at B. east of C. due south of D. west of

11. The earth is classified as a 11.____
A. star B. comet C. satellite D. planet

12. One effect among the following of the earth's rotation on its axis is the 12.____
A. daily westward shift of stars
B. variation in the seasons
C. moon's phases
D. formation of circular star trails in the polar region

13. The planet whose solid surface can be observed is 13.____
A. Venus B. Mars C. Saturn D. Jupiter

14. The phase of the moon at which it rises CLOSEST to noontime is 14.____
A. new moon B. full moon
C. first quarter D. last quarter

15. The surface temperature of a blue-white star is CLOSEST, in degrees F, to 15.____
A. 11,000 B. 20,000 C. 28,000 D. 36,000

16. The MAXIMUM possible duration of a total lunar eclipse is CLOSEST to 16.____
A. 15 minutes B. 30 minutes C. 1 hour D. 2 hours

17. A satellite with an average distance of 300 miles above the earth's surface would revolve around the earth in a period CLOSEST to 17.____
A. 50 minutes B. 90 minutes C. 10 hours D. 24 hours

18. Of the following, the one generally credited with sponsoring the heliocentric theory of the solar system is 18.____
A. Aristotle B. Copernicus
C. Hippocrates D. Ptolemy

19. Karl Jansky was the discoverer of 19.____
A. a naked-eye comet in 1958
B. a new moon of Jupiter
C. radio waves from outer space
D. the F layer of the ionosphere

20. According to modern concepts of the life history of a star, a dwarf star represents the stage of star 20.____
A. birth B. youth C. maturity D. old age

21. If Star A is a 5th magnitude star and Star B is a 2nd magnitude star, then the ratio of the brightness of B to that of A is CLOSEST to 21.____
A. 2.5 B. 3.0 C. 9.0 D. 15.0

22. Which one of the following is a first magnitude star in the constellation Orion? 22.____
A. Altair B. Aldebaran C. Betelgeuse D. Sirius

23. Earthshine is MOST prominent at the phase of the moon called 23.___
    A. new crescent B. first quarter
    C. full moon D. old gibbous

24. The moon's speed of revolution is 24.___
    A. greatest at apogee
    B. greatest at perigee
    C. greatest at new moon phase
    D. uniform throughout its orbit

25. Quantitative analysis of the stars through spectrographic study indicates that the stars consist of APPROXIMATELY 25.___
    A. 50% uranium
    B. 75% hydrogen
    C. 20% thorium
    D. the same percentage composition of elements as the earth

---

KEY (CORRECT ANSWERS)

1. B
2. C
3. A
4. D
5. D

6. C
7. A
8. C
9. C
10. D

11. D
12. D
13. B
14. C
15. C

16. D
17. B
18. B
19. C
20. D

21. D
22. C
23. A
24. B
25. B

TEST 4

DIRECTIONS: Each question or incomplete statement is followed by several suggested answers or completions. Select the one that BEST answers the question or completes the statement. *PRINT THE LETTER OF THE CORRECT ANSWER IN THE SPACE AT THE RIGHT.*

1. The manner of occultation of a planet by the moon shows that the moon demonstrates which one of the following characteristics? 1.____
   A. Lacks an atmosphere
   B. Moves faster than the planet in miles per hour
   C. Rotates on its axis
   D. Frequently changes its axis of rotation

2. The moon moves MOST rapidly in its orbit at 2.____
   A. apogee B. full moon C. new moon D. perigee

3. Mercury has a low albedo because of its 3.____
   A. high temperature B. small size
   C. small mass D. lack of atmosphere

4. Astronomers have found that, as viewed from the earth, 4.____
   A. all galaxies are receding at the same speed
   B. some galaxies are approaching the earth
   C. the more distant the galaxy, the greater its speed of recession
   D. the more distant the galaxy, the less its speed of recession

5. In astronomy, the measure which corresponds to longitude is called 5.____
   A. declination B. right ascension
   C. left ascension D. zenith variation

6. The theory of the expanding universe of the stars is based largely on the evidence that analysis of light coming from the distant stars shows 6.____
   A. a decrease in wave-length
   B. a shift toward the violet end of the spectrum
   C. no shift in Fraunhofer lines
   D. a shift toward the longer wave-lengths

7. Telstar went into orbit with a speed of approximately 18,000 miles per hour at perigee, which is about 600 miles. If apogee is approximately 3500 miles, its speed at apogee should be, in miles per hour, CLOSEST to which one of the following? 7.____
   A. 6,000 B. 11,000 C. 16,000 D. 18,000

8. The study of tidal effects and careful measurement have led to which one of the following conclusions? 8.____
   A. The period of the earth's rotation is gradually decreasing.
   B. Tidal forces are greater on earth than they are on the moon.
   C. The moon is slowly spiraling outward from the earth.
   D. The moon's period of rotation is decreasing.

9. The ecliptic is ALWAYS inclined 23.5° to the 9.____
   A. celestial equator B. horizon
   C. Tropic of Cancer D. diurnal circle

10. As a planet revolves about the sun, its speed 10.____
   A. is constant
   B. increases as it approaches the sun
   C. decreases as it gets closer to the sun
   D. varies so that all planets complete one revolution in 365 days

11. The ratio of the brightness of a star of magnitude 3 to that of a star of magnitude 4 is 11.____
   A. 2:5 B. 3:4 C. 4:3 D. 5:2

12. All of the following statements regarding a solar eclipse are true EXCEPT that 12.____
   A. it can occur only once in any calendar year
   B. it can occur only at new moon phase
   C. at any one point on the earth, the maximum duration of a total eclipse is about 7 minutes
   D. the maximum width of the moon's umbra on the earth is approximately 167 miles

13. The ratio of the distance from the sun of a theoretical solar system planet with a period of revolution of eight earth years to the earth's distance from the sun is CLOSEST to which one of the following? 13.____
   A. 2:1 B. $2\sqrt{2}$:1 C. 4:1 D. 8:1

14. As their distance from the earth changes, planets 14.____
   A. change in brightness and apparent size
   B. change in brightness but not in apparent size
   C. change in apparent size but not in brightness
   D. do not change in brightness or apparent size

15. Which one of the following is known for having obtained evidence of the rotation of the earth? 15.____
   A. Archimedes B. Foucault C. Maury D. Wright

16. The gravitational force between two bodies in space varies 16.____
   A. *directly* with the distance between them
   B. *inversely* with the distance between them
   C. *inversely* with the square of the distance between them
   D. *directly* with the square of the distance between them

17. If the inclination of the earth's axis were 45°, the summers in New York City would be 17.____
   A. cooler   B. warmer   C. shorter   D. longer

18. Of the following, the one MOST closely associated with the production of solar energy is 18.____
   A. burning of hydrogen gas   B. fusion of hydrogen nuclei
   C. fission of U^{235} nuclei   D. fission of helium nuclei

19. During which phase of the moon is an eclipse of the sun MOST likely to occur? 19.____
   A. New   B. Full   C. Quarter   D. Gibbous

20. Which one of the following scientists discovered laws governing the motion of planets? 20.____
   A. Roentgen   B. Wöhler   C. Foucault   D. Kepler

21. Circular star trails seen in photographs are an effect of 21.____
   A. the revolution of the earth about the sun
   B. cosmic radiation
   C. the rotation of the earth
   D. an expanding universe

22. Of the following, the constellation NOT visible in the Northern Hemisphere in the winter sky is 22.____
   A. Lyra   B. Cassiopeia   C. Ursa Major   D. Ursa Minor

23. The rays of the sun are vertical over the Tropic of Capricorn at a time CLOSEST to which one of the following dates? 23.____
   A. March 21   B. June 21   C. September 21   D. December 21

24. Moonrise occurs CLOSEST to solar noon at the phase of the moon known as 24.____
   A. first quarter   B. full moon
   C. third quarter   D. new moon

25. The bright star Capella is in the constellation 25.____
   A. Auriga   B. Bootes   C. Lyra   D. Scorpius

KEY (CORRECT ANSWERS)

1. A
2. D
3. D
4. C
5. B

6. D
7. B
8. C
9. A
10. B

11. A
12. A
13. C
14. A
15. B

16. C
17. B
18. B
19. A
20. D

21. C
22. A
23. D
24. A
25. A

EXAMINATION SECTION

TEST 1

DIRECTIONS: Each question or incomplete statement is followed by several suggested answers or completions. Select the one that BEST answers the question or completes the statement. *PRINT THE LETTER OF THE CORRECT ANSWER IN THE SPACE AT THE RIGHT.*

1. The synodical month is 1.____
   A. about two days longer than the sidereal month
   B. about three days shorter than the sidereal month
   C. 31 days long on the average
   D. the time required for the moon to revolve 360°

2. Betelgeuse and Rigel are navigational stars in the constellation 2.____
   A. Cassiopeia B. Lyra C. Orion D. Ursa Major

3. Which one of the following statements about the planet Jupiter is NOT true? 3.____
   A. All of its moons revolve from west to east.
   B. Its rotation period is shorter than any other planet.
   C. It has more natural moons than any other planet.
   D. Some of Jupiter's moons are larger than the planet Mercury.

4. The principal reason that an eclipse of the moon does NOT occur every full moon is that the 4.____
   A. difference between perigee and apogee is too large
   B. refraction of sunlight interferes
   C. earth's shadow barely reaches the moon
   D. moon's orbit is inclined to the plane of the earth's orbit

5. The ratio of the diameter of the sun to that of Jupiter is CLOSEST to which one of the following? 5.____
   A. 5:1 B. 10:1 C. 25:1 D. 100:1

6. Of the following gases, the one MOST prominent in the atmosphere of Jupiter is 6.____
   A. acetylene B. methane
   C. nitric oxide D. sulfur dioxide

7. At the level of tropopause, the MOST abundant gas in the atmosphere is 7.____
   A. hydrogen B. helium C. oxygen D. nitrogen

8. The planet with the GREATEST orbital speed is 8.____
   A. Earth B. Mercury C. Neptune D. Pluto

9. Of the following, the FIRST scientist to use the astronomical telescope was 9.____
   A. Galileo B. Kepler C. Leeuwenhoek D. Tycho Brahe

10. In an annular eclipse of the sun, 10.___
   A. both the umbra and penumbra of the moon touch the earth
   B. the moon's umbra touches the earth, but the penumbra does not
   C. the moon's penumbra touches the earth, but the umbra does not
   D. neither the umbra nor the penumbra of the moon touches the earth

11. Of the following, the one which is an example of a diffuse nebula is the 11.___
   A. Great Cluster in Hercules B. Great Nebula in Orion
   C. Great Spiral in Andromeda D. Milky Way

12. The atmosphere of which one of the following planets is rich in methane and ammonia? 12.___
   A. Venus B. Mercury C. Jupiter D. Mars

13. The ratio of the volume of the earth to that of the moon is CLOSEST to which one of the following? 13.___
   A. 4:1 B. 8:1 C. 32:1 D. 64:1

14. Which one of the following statements is TRUE of the stars of the same spectral class? They 14.___
   A. are all equidistant from the sun
   B. are all blue-white stars
   C. are all Cepheid variables
   D. all have approximately the same surface temperature

15. Of the following, the diameter of our galaxy in light-years is CLOSEST to 15.___
   A. 1,000 B. 10,000 C. 100,000 D. 1,000,000

16. According to modern astronomical theory, the average life-time for a star is CLOSEST, in years, to which one of the following? 16.___
   A. 10 million B. 100 million C. 1 billion D. 10 billion

17. Of the following, the layer of the sun whose thickness is SMALLEST is the 17.___
   A. chromosphere B. corona
   C. photosphere D. reversing layer

18. Rigel is a navigational star in which one of the following constellations? 18.___
   A. Orion B. Canis Major C. Cassiopeia D. Ursa Major

19. Which one of the following constellations is MOST commonly used as an aid in finding the North Star? 19.___
   A. Orion B. Pleiades C. Taurus D. Ursa Major

20. Which one of the following planets is 1.52 astronomical units from the sun? 20.___
   A. Saturn B. Mars C. Venus D. Mercury

21. Cepheid variable stars go through a light cycle which, at its maximum, may be as long as 21.____
A. 1 hour B. 1 day
C. several weeks D. several years

22. Astronomers use the term *magnitude* to denote 22.____
A. size of stars
B. apparent brightness
C. telescope range
D. telescope magnification ratio

23. When the earth is farthest from the sun, the season in the earth's Southern Hemisphere is 23.____
A. spring B. summer C. fall D. winter

24. The interval of 18 years, 11 1/3 days after which a cycle of solar and lunar eclipses is repeated, is known as the 24.____
A. ecliptic limit B. eclipse season
C. saros D. albedo

25. The average length of the synodic month is CLOSEST to which one of the following? _____ days. 25.____
A. 27.5 B. 29.5 C. 31.5 D. 33.5

KEY (CORRECT ANSWERS)

1. A
2. C
3. A
4. D
5. B
6. B
7. D
8. B
9. A
10. C
11. B
12. C
13. D
14. D
15. C
16. D
17. C
18. A
19. D
20. B
21. C
22. B
23. D
24. C
25. B

TEST 2

DIRECTIONS: Each question or incomplete statement is followed by several suggested answers or completions. Select the one that BEST answers the question or completes the statement. *PRINT THE LETTER OF THE CORRECT ANSWER IN THE SPACE AT THE RIGHT.*

1. The planet in our solar system NEAREST in size to the earth is 1.____
   A. Mars B. Mercury C. Neptune D. Venus

2. Of the following, a star of the SECOND apparent visual magnitude is the star 2.____
   A. Arcturus B. Polaris C. Procyon D. Rigel

3. The famous Horsehead Nebula in Orion is a 3.____
   A. cluster of stars B. dark nebula
   C. galaxy D. Magellanic Cloud

4. The planet in our solar system with the MOST eccentric orbit is 4.____
   A. Jupiter B. Mercury C. Pluto D. Uranus

5. The great telescope at Jodrell Bank, England, is a 5.____
   A. radiotelescope B. prism
   C. refractor D. Schmidt telescope

6. Astronomers use spectroscopes to 6.____
   A. measure the distance to stars
   B. determine the mass of stars
   C. determine the size of stars
   D. analyze the light from stars

7. The planet with the LARGEST number of moons is 7.____
   A. Mars B. Jupiter C. Saturn D. Uranus

8. The Milky Way is a 8.____
   A. galaxy B. constellation
   C. spiral nebula D. solar system

9. Of the following, the planet that shows phases when viewed through the telescope is 9.____
   A. Mars B. Venus C. Saturn D. Jupiter

10. Of the following, the star of the second apparent magnitude is 10.____
    A. Altair B. Aldebaran C. Polaris D. Rigel

11. Of the following, a star in the constellation Lyra is 11.____
    A. Antares B. Capella C. Sirius D. Vega

12. Our sun is classed as a(n) _____ star. 12.___
A. average-size B. dwarf
C. giant D. super-giant

13. Man-made satellites travelling about the earth in elliptical orbits are CLOSEST to the earth at their 13.___
A. nadir B. perigee C. apogee D. zenith

14. A planet which can NEVER be seen from the earth at midnight is 14.___
A. Mars B. Venus C. Jupiter D. Saturn

15. The orbits of the comets are USUALLY 15.___
A. circles B. ellipses C. parabolas D. hyperbolas

16. The SECOND largest planet in our solar system is 16.___
A. Jupiter B. Neptune C. Uranus D. Saturn

17. A celestial object which is NOT a part of our galaxy is the 17.___
A. Great Spiral Nebula in Andromeda
B. Great Nebula in Orion
C. Great Cluster in Hercules
D. Omega Centauri

18. A star whose apparent visual magnitude is expressed as a negative number is 18.___
A. Algol B. Betelgeuse C. Spica D. Sirius

19. The earth revolves around the sun at a speed, in miles per hour, of about 19.___
A. 25,000 B. 66,000 C. 93,000 D. 186,000

20. The planet with the SHORTEST period of rotation is 20.___
A. Earth B. Mars C. Mercury D. Jupiter

21. When Mercury and Venus are morning stars, 21.___
A. both are seen near the eastern horizon
B. both are seen near the western horizon
C. Mercury is in the east, Venus in the west
D. Mercury is in the west, Venus in the east

22. The dark side of the moon cannot be observed from the earth because the moon 22.___
A. revolves around the earth, but does not rotate on its axis
B. rotates about its axis, but does not revolve around the earth
C. has the same period of rotation as its period of revolution
D. has a period of rotation of 24 hours and a period of revolution of 30 days

23. The SMALLEST of the planets in the solar system is 23.___
A. Mars B. Uranus C. Pluto D. Mercury

24. An astronomical feature which lies beyond the limits of the Milky Way is 24.___
   A. the Pleiades cluster
   B. the Perseids
   C. Halley's Comet
   D. the Great Nebula in Andromeda

25. The famous Horsehead Nebula is in the constellation 25.___
   A. Canis Major B. Orion C. Taurus D. Sagittarius

KEY (CORRECT ANSWERS)

1. D
2. B
3. B
4. C
5. A

6. D
7. B
8. A
9. B
10. C

11. D
12. A
13. B
14. B
15. B

16. D
17. A
18. D
19. B
20. D

21. A
22. C
23. D
24. D
25. B

TEST 3

DIRECTIONS: Each question or incomplete statement is followed by several suggested answers or completions. Select the one that BEST answers the question or completes the statement. *PRINT THE LETTER OF THE CORRECT ANSWER IN THE SPACE AT THE RIGHT.*

1. The phase of the moon which is viewed when the moon sets shortly after the sun is called the 1.____
   A. new crescent B. first quarter
   C. new gibbous D. last quarter

2. The Perseids are a 2.____
   A. meteor swarm
   B. group of planetoids
   C. group of sunspots
   D. star cluster in the constellation Perseus

3. Auroras or northern lights are believed to be caused by 3.____
   A. reflection of sunlight from the polar icecap
   B. streams of electrically-charged particles deflected and concentrated by the earth's magnetic field
   C. diffraction of sunlight by ice crystals in the upper atmosphere
   D. disintegration of radioactive elements in the earth's atmosphere

4. Neap tides occur when there is a 4.____
   A. full moon
   B. new moon
   C. lunar eclipse
   D. moon at first quarter phase

5. When an artificial satellite orbits the earth, its position CLOSEST to the earth is known as 5.____
   A. perigee B. aphelion C. apogee D. perihelion

6. Of the following, the statement which is NOT true is that the sun 6.____
   A. is a star
   B. is a stationary body
   C. varies in its distance from the earth
   D. rotates on its axis

7. The earth is FARTHEST from the sun in 7.____
   A. January B. March C. July D. October

8. The constellation that can be seen at any time of the year in New York is 8.____
   A. Cassiopeia B. Orion C. Andromeda D. Leo

9. A star in the constellation Orion is 9.____
A. Antares B. Arcturus C. Capella D. Rigel

10. Two stars differ in apparent brightness by exactly one magnitude. 10.____
The ratio of their apparent brightness is, therefore,
A. 1:1 B. 1:2.5 C. 1:5.0 D. 1:10

11. In light-gathering power, a 200" reflecting telescope is GREATER than a 100" reflecting telescope by a ratio of 11.____
A. 2:1 B. 4:1 C. 16:1 D. 100:1

12. A wide-angle type of reflecting telescope especially adapted to photographing large areas of the heavens is known as the 12.____
A. Balmer B. Cassegrainian
C. Newtonian D. Schmidt

13. The ratio of the light reflected by a planet to that received on its whole illuminated hemisphere is called its 13.____
A. albedo B. insolation
C. luminosity D. noctilucence

14. The shift of the red lines in the spectrum of light reaching us from distant galaxies (Doppler effect) makes it possible to determine their 14.____
A. temperatures
B. velocities and directions of movement
C. sizes and shapes
D. chemical compositions

15. The planet in our solar system which has the MOST eccentric orbit around the sun is 15.____
A. Uranus B. Pluto C. Neptune D. Saturn

16. To an observer on Earth, the BRIGHTEST planet is 16.____
A. Jupiter B. Saturn C. Mars D. Venus

17. The Russian Lunik revolves around the 17.____
A. sun outside the earth's orbit
B. sun inside the earth's orbit
C. moon
D. earth

18. The Northern Cross lies in the constellation 18.____
A. Cygnus B. Bootes C. Lyra D. Pegasus

19. A galaxy visible to the unaided eye lies in the constellation 19.____
A. Andromeda B. Ursa Minor C. Auriga D. Canis Major

20. Spring tides occur at 20.____
A. full moon only B. new moon only
C. both full and new moon D. first and last quarter phases

21. An annular eclipse of the sun takes place at the phase of the moon called 21.___
   A. new moon   B. new gibbous
   C. new crescent   D. full moon

22. At perigee, our moon's distance, expressed in miles, from the earth is about 22.___
   A. 205,000   B. 220,000   C. 235,000   D. 245,000

23. Moore and Ross recently ascended to a height of 80,000 ft. and discovered 23.___
   A. cosmic ray bombardment from space
   B. the nature of the rings of Saturn
   C. water vapor in the atmosphere of Venus
   D. new values for the space unit Mach

24. All stars visible to the naked eye belong to 24.___
   A. the solar system
   B. a number of galaxies
   C. the Milky Way galaxy
   D. the Milky Way galaxy and several spiral nebulae

25. The BRIGHTEST star visible in the nighttime sky in New York City is 25.___
   A. Betelgeuse   B. Orion   C. Polaris   D. Sirius

KEY (CORRECT ANSWERS)

1. A
2. A
3. B
4. D
5. A

6. B
7. C
8. A
9. D
10. B

11. B
12. D
13. A
14. B
15. B

16. D
17. A
18. A
19. A
20. C

21. A
22. B
23. C
24. C
25. D

BASIC FUNDAMENTALS OF ASTRONOMY

CHAPTER 1

THE UNIVERSE

Astronomers, who study the stars, tell us that we live in a universe filled with millions and millions of stars like our sun. The universe is so large that it takes light thousands of millions of years to move the distances we can see in our telescopes. How did the astronomers learn about the stars and the great distances in space? This is the question we shall try to answer in this chapter.

Radiation

Every star, like our sun, is extremely hot inside. A star is a nuclear furnace, busy changing the element hydrogen into the element helium. This process produces vast amounts of energy. If you go to a beach in the summertime and your skin tans, you have used some of that energy. Every morsel of food you eat has been grown by the sun's energy. How does this energy reach the earth? It comes in the same way a radio broadcast reaches your radio. It is sent through space as radiation.

A radio station broadcasts on one frequency. We must tune to that frequency to hear the station. The stars broadcast their radiation on many frequencies. One of these frequencies - called the ultraviolet range - is tuned in by your skin. The energy on this frequency is used to tan the skin. Some of the frequencies coming from the stars can be seen by human eyes. We call these frequencies *light*. Others are like radio waves - they cannot be seen. The whole universe is filled with these radiations coming from the stars.

Light

On most clear nights we can see the moon in the sky. But the moon does not shine or radiate. It is a dead body. We can see the moon only because light from the sun is reflected from it to earth. In fact, the only reason we can see anything at all is that light reflects into our eyes from different objects in different ways. We have learned how to interpret these reflections. We say that we *see* things. Actually, we see the light that reflects off things and reaches our eyes. When there is no light, we cannot see things at all. But they are still there. That is why we bump into solid objects when we are walking in the dark.

Because sight is very important to us and we see only if there is light, scientists have studied it very carefully. They have learned and put to use a number of facts about light. Here are five of the most important characteristics of light that scientists use.

First, *light moves at a constant speed*. When we snap on a light switch, the light floods the room instantly. Or so it seems to us because it moves so fast. Scientists have been able to measure the speed of light. They find that it moves at a speed of about 186,000 miles per second. Radio waves and electricity travel at this same rate of speed. In fact, scientists believe that nothing else in the universe travels faster than light. Because light moves at a steady speed, it can be used for measuring. The astronomer uses the *light year* as a measuring standard. This is the distance light travels in one year at a speed of 186,000 miles per second. We could not use the speed of light as a standard if light did not move at a constant speed.

Second, *light moves in a straight line*. When a beam of light strikes any solid object that is not transparent, it cannot move through it. The object casts a clear black shadow. Light cannot bend around corners. That is why we cannot see around corners. We can hear around corners because sound travels in air. And sound waves do move around corners. But light travels only in a straight line. We cannot see unless there is a straight clear path to the thing we are trying to look at.

Third, *light can be reflected*. We can see the moon because it reflects light from the sun. We can see ourselves in a mirror for the same reason. In fact, if light were not reflected, we could see nothing at all. For everything we see, except light itself - the sun and stars and light bulbs and fire - we see by reflected light. This ability to reflect light is made use of in a special kind of telescope, called a *reflecting telescope*.

Fourth, *light can be bent*. Light cannot move around a corner, but it can be bent. If you place a stick in a pool of water, the stick seems to bend where it touches the water. The light reflected from the stick is bent by the water. This is called *refraction*. Every telescope and microscope uses this principle. So do eyeglasses. They bend the rays of light and focus them in the right place. You can see how this works by looking at the figure. The ground glass bends the light rays. If it is ground properly, all of the rays are bent the same amount. This makes a clear picture which is larger and easier to see.

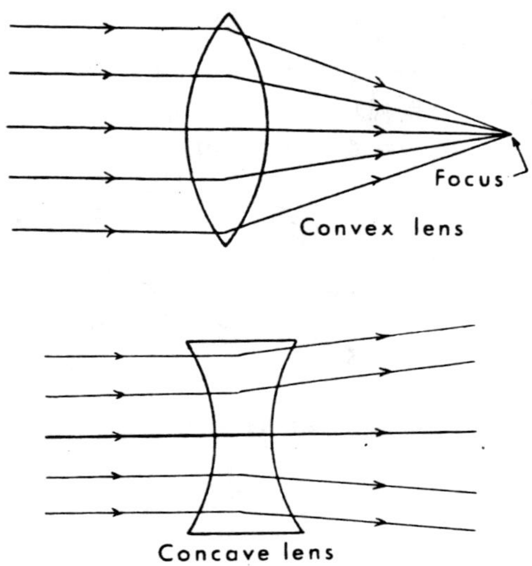

FIGURE Refraction of Light Through a Lens.

Fifth, *light is a series of frequencies*. Light radiates as a mixture of several frequencies, not as just one wave. Each frequency appears to our eyes as a separate color. Light is like a dozen radio stations all broadcasting on different frequencies. When we *see*, we tune in all of these frequencies at the same time. When we see one color, we see one of these frequencies. The others have been lost somewhere. We will learn more about this when we study color.

These five characteristics of light are used by scientists to study the universe.

The Optical Telescope

One way that astronomers have learned about the universe is by studying the radiations coming from the stars. The frequency they study most is light. Light can be studied through optical telescopes.

A telescope is designed to capture light from distant stars and concentrate it into a small area that can be seen and understood. Our eyes are very small, and only a little light reaches them at one time. The telescope gathers light from a large area and bends the light rays until they fit the small area the eye can see. This gives a brighter and stronger picture.

An optical telescope can use a mirror to gather light, or it may use a curved piece of glass called a *lens*. The amount of light that the telescope gathers depends upon the size of the mirror or lens. A telescope that uses a mirror to gather light is called a *reflecting* telescope. The mirror reflects the light that it gathers. A telescope that uses a lens to gather light is called a *refracting* telescope. The lens refracts, or bends, the light it gathers. A large mirror or lens gathers more light than a small one. Because a big mirror can be made more easily than a big lens, our largest telescopes are reflecting telescopes. The largest reflecting telescope in the world has a mirror two hundred inches in diameter. It is located on Mount Palomar in California. The lens of the largest refracting telescope is forty inches in diameter. It is located at Yerkes Observatory on Lake Geneva.

The Refracting Telescope

The small hand telescope is usually a refracting telescope. It uses a lens to gather light. The figure below shows a typical

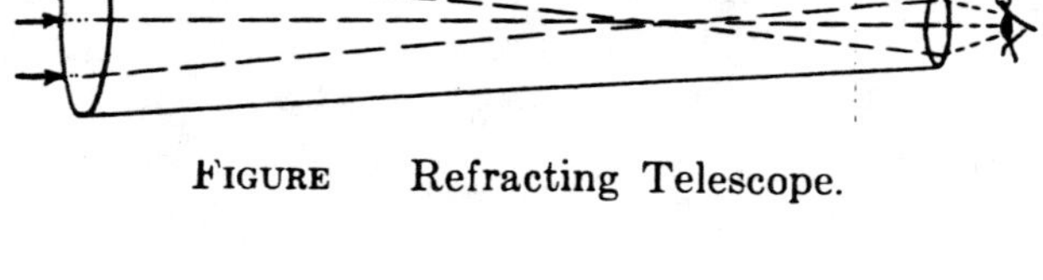

FIGURE Refracting Telescope.

refracting telescope. Light enters the telescope through the lens in front. It is bent, or refracted, by the lens. This pulls the light together into a smaller pattern. The light rays cross over in the middle of the telescope. Then they strike the eyepiece and produce a smaller picture of the light entering the lens. This small picture can be seen by the eye. Because the light rays will have crossed, the picture as seen by the eye will be inverted, or upside-down.

By placing a camera at the end of a telescope, instead of one's eye, astronomers are able to photograph the heavens.

Reflecting Telescopes

The lens in a refracting telescope must be ground perfectly smooth to prevent distortion. This cannot be done with very large lenses. The largest telescopes, therefore, must use the curved mirror to gather light. The light is reflected off the curved surface of the mirror and focused into an eyepiece.

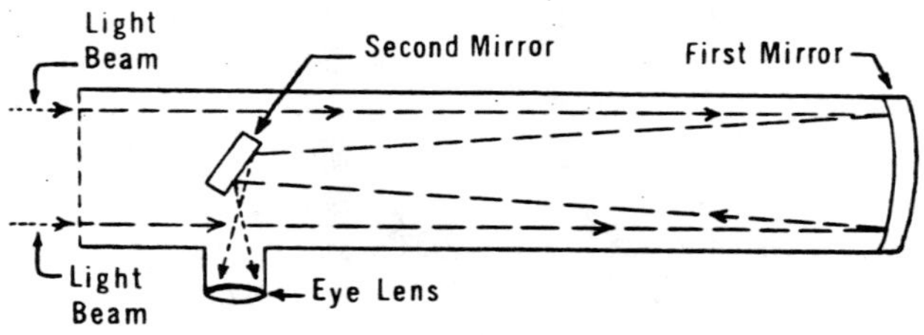

Using Telescopes

The amount of light gathered by a telescope depends upon the size of the lens or mirror it uses. A telescope with a thirty-six inch lens gathers about thirty thousand times as much light as the human eye. This shows the astronomer many more stars than his eye can see unaided. On a clear night you can look up and see a few thousand stars in the sky. But with a four-inch telescope you can see more than two million stars. The giant reflecting telescope on Mount Palomar shows billions of stars that are too far away to be seen by the naked eye.

The larger telescopes are able to separate the light coming from the different stars and give a clear picture of each one. This is called *resolving power*. It is very important to astronomers. One of the meanings of *resolve* is to *separate or break up into elements*. A cluster of stars may look like a blur of light in a small telescope. In a larger scope the blur of light is seen as many separate stars. The elements of the blur of light - the stars, that is - are resolved by the stronger telescope.

Thermocouple

Astronomers use the telescope more than any other tool available. But they also make use of other instruments. We have seen that the camera has helped astronomers make a record of the heavens. An instrument called the *thermocouple* is used with a telescope to measure the temperature of distant stars.

The Spectroscope

Another instrument, called the *spectroscope*, is used to show the type of material in different stars. Light from a star can be beamed into the spectroscope through any telescope. The light passes through a triangle of glass, called a *prism*. This prism bends the light rays. It bends every frequency a different amount. Thus it has the effect of separating the different frequencies of light coming from the star. Each frequency appears as a band of color in the spectroscope.

By learning the patterns of frequencies produced by different materials, the scientist has learned how to interpret these lines. From spectroscopic studies, we know that the sun is composed of many of the same elements we have on earth. In this way we have also learned the composition of many stars far away in the universe. We have found out that most stars are made of some of the same types of material we find here on earth.

The Radio Telescope

Radiation from the stars includes many frequencies that our eyes cannot see. In recent years scientists have begun to study these invisible radiations to learn more about the universe. The instrument they use is called a *radio telescope*. It is really a very powerful radio receiver with a special antenna. The signals received from space sound much like static on a regular radio. In fact, radio astronomy was discovered because of static caused by radiation from space. The astronomers have learned how to *tune in* broadcasts from outer space. Because the signals are very weak, a large antenna is used to strengthen them.

Inside the radio telescope, the faint electrical signals coming from space are *amplified*, or strengthened, thousands of times. The signals received by the telescope are recorded to give a permanent record. This record can be studied. When the radio telescopes pick up a source of large electrical radiations, scientists try to find the area with a large optical telescope. Sometimes the cause of the radiation can be seen. If so, the astronomers can interpret the signals they receive by radio. One astronomer has already taken a picture that shows two huge galaxies colliding or passing through one another. The collision produced very strong radiation signals. These signals were detected by a radio telescope. The astronomer located the area with the two-hundred-inch optical telescope and took the picture.

The radio telescope provides many advantages, once we learn to use it. It can *see* further into the universe. It can be used in the daytime. It is not bothered by clouds. Used with the optical telescope, the radio telescope is rapidly expanding our knowledge of the vast universe around us.

CHAPTER 2

STARS AND GALAXIES

Bodies in Space

Can you imagine a single star so large that it could hold a thousand million bodies the size of the earth? There are stars in the universe as large as that. We have nothing that we can compare them to. This is one of the problems we meet when we study the whole universe. Everything is so large and things move so fast that it is hard to comprehend them. Here on our planet a quart jar of earth contains billions of billions of billions of atoms. In some parts of the universe, a quart jar would contain only one tiny atom. How can we imagine space that holds absolutely nothing? Space around the earth contains atmosphere. We cannot see it or feel it, but it is there all the same. Large areas in outer space contain almost nothing.

Even in the part of the universe we can see, the earth is only a tiny speck. Our sun is only one very ordinary little star among the millions of stars in the Milky Way. Most of the other stars in this galaxy are about the same size as the earth. Some are larger than the sun. Others are smaller than the earth.

Far out in space there are millions of other galaxies, some larger and some smaller than the Milky Way. We do not even know how many galaxies there are. Most astronomers believe that there are many millions of planets throughout the universe. But no one has seen any outside our own solar system.

Galaxies

Most of the matter in the universe is gathered together in great clusters called galaxies. Besides its millions of stars, the Milky Way galaxy also contains probably millions of planets. In addition, there are immense clouds of dust and gas spread around the galaxy. In fact, about half of all the matter in the universe exists in the form of dust and gas. Many astronomers believe that new stars, like the sun, are formed from these great clouds.

The galaxies are very large. It takes light hundreds of thousands of years to pass from one end of a galaxy to the other. Since the distance between galaxies is measured in millions of light years, there is very little danger of a crash. This is a good thing, because the galaxies move through space at terrific speeds.

Radiation

Every visible star in the universe produces radiation. Great amounts of energy are thrown into space by the atomic furnace inside each star. This radiation travels through space at the speed of light.

Every part of the universe is filled with this radiant energy from stars. Planets like the earth catch and hold some of the energy from this radiation. We are able to live because the earth can capture from the sun this radiant energy which provides heat and plants for food.

Maybe other planets contain living things that use radiant energy to live, too. Since there are millions of stars, it is probable that many of them have planets similar to the earth. Far away in space, perhaps millions of light years from earth, another planet may hold people who look at our sun through their telescopes. We may never know whether they do for certain, unless they are fairly close to earth. No man could live long enough to make the journey.

Stars

The stars are the queens of the universe. The great heat and pressure inside the star raises the temperature to several million degrees centigrade. Atomic fusion takes place at this high temperature. Stars consume hydrogen nuclei, or use hydrogen nuclei to produce atomic energy. When four hydrogen nuclei are combined to form a helium nucleus, energy is released. This energy is radiated into space by the star. When the supply of hydrogen in a star is gone, we say the star dies. Our own sun will stop consuming all of its hydrogen some day. But no one needs to worry about that for a long time. There is enough hydrogen in the sun to last for millions of years.

The sun is a typical star. There are billions of other stars of roughly the same size and temperature. These stars are called *main sequence* stars. The sun is about 860,000 miles in diameter - just about average for main sequence stars. Inside, the temperature of the sun is about 20 million degrees C. Its surface temperature is much lower - about 6,000°C. About 99 percent of all the stars we can see are main sequence stars.

There are two other types of stars in the universe. One type is much larger than the sun, the other is smaller. The larger stars are called *red giants*. They are the real giants of the universe. These stars are cooler than the sun and radiate at a different frequency. Their radiation appears red in the spectroscope. That is why they are called red giants. Their size is really amazing. One red giant, called *Betelgeuse* (Bē'tel jōōz), is 25 million times as large as our sun.

The smaller stars are called *white dwarf* stars. They are much smaller than the sun and hotter. They are about the size of the earth. But the matter in white dwarfs is packed so tight that a cubic inch of the stuff has as much mass as a freight car full of coal. And they are exceedingly hot.

Inside A Galaxy

The large galaxy called the Milky Way, where we live, looks something like the one shown in the figure below. The galaxy in the

picture looks like a giant pinwheel in a firework display. Not all galaxies are shaped this way. Some are a round cluster of stars. Some are shaped something like an H. The large galaxies, like the Milky Way and *Andromeda*, are spiral galaxies. In a spiral galaxy the stars are grouped together in the long curved arms of the spiral. There are several million stars in just this one galaxy. Our galaxy covers the distance of a hundred thousand light years from one edge to the other.

Scientists do not know exactly how many stars there are altogether in the universe. One astronomer estimated that there were 10,000,000,000,000,000,000,000. There is no way that we can compare this number on earth. The galaxies are usually spinning as they speed through space.

All galaxies that we can see are filled with great clouds of dust and gas. Many astronomers believe that new stars are born inside these clouds of dust. We know that some stars are young and others are old;

so we are certain that the stars were not all produced at the same time. A star lives by *burning* hydrogen and producing helium. By measuring the amount of helium, we can tell how long a star has been burning. Some stars are much older than others. In fact, some stars have probably burned out long ago. Others have gathered themselves together for one last effort and exploded.

How can a cloud of gas form a new star? The answer lies in the movement of the cloud of dust and gas and in Newton's law of gravitation. The movement of the gas and dust creates powerful currents inside the cloud. This tends to bring particles together. The area where the particles are crowded together then has a larger mass than other parts of the dust cloud. According to Newton's law, the attractive force of a body depends upon its mass. Slowly, the mass of gas and dust grows larger and heavier. The larger and heavier it grows, the more its mass increases. This attracts still more gas and dust.

These clouds are extremely large. Some are millions of millions of miles across. The rotation of the cloud mass tends to pull it into the shape of a ball. The dust and gas at the outer edge of the ball are drawn toward the center by the pull of gravity, thus increasing the pressure on the particles inside the rotating ball. The increase in pressure increases the mass and creates an even greater pull from the outside toward the center. This causes the temperature to increase. When the temperature gets high enough, an atomic reaction occurs. The cloud, now a solid mass of tightly packed dust and gas, begins to burn. A new star has been born.

This explanation is only a theory of how stars are made. No one has ever seen a star made this way. The process probably takes thousands or millions of years. It does seem to explain the facts that we know at this time. This theory is accepted by many scientists as the best explanation we have for star making.

Explosions

When a star uses up its supply of energy, it stops burning. Some stars go out in a blaze of glory. They pull themselves together and explode with a tremendous bang. The figure below shows an explosion

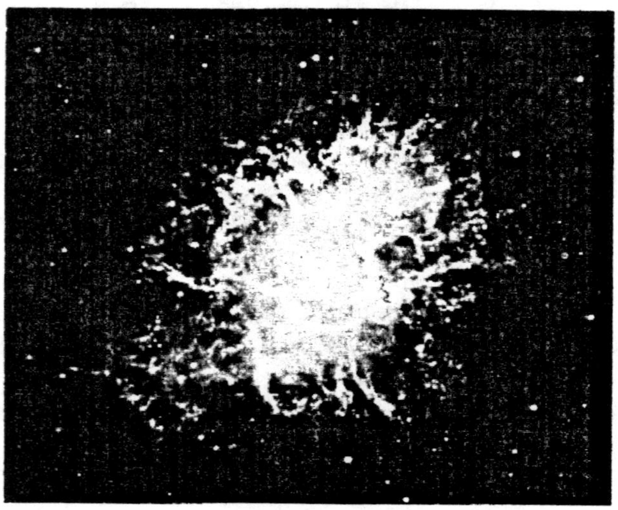

that took place more than nine hundred years ago. The gases and matter are traveling from the explosion at a rate of six hundred miles per second. You can imagine how far they have traveled in nine hundred years. Yet the explosion still looks small in our telescopes. This will give you an idea of how far away the explosion took place.

When a star explodes, the explosion is called a *nova*. It is not unusual in the universe, where there are billions of stars. Some of them stop burning and explode every few years. Once in awhile, however, there is a real super explosion called a *super nova*. The figure shows one of these extra large explosions. Scientists know about a half dozen of them in the last thousand years. The energy that is released in one of those explosions is so great that to compare it with a hydrogen bomb would be like comparing the sun with a tiny spark.

Most scientists do not believe that our sun will explode. They think it will just fade out and stop burning. It may flare up strongly before it stops burning. If it does, the earth will be burned to a crisp before it turns into a cold solid ball.

Questions About The Universe

There are many questions about the universe that we would like to answer. How old is it? Where did it come from? Is the universe slowly thinning out, or is it growing more dense? As yet, scientists cannot answer such questions. Perhaps they may never be able to.

We try to estimate the age of stars by measuring the helium they contain. But each year our estimates grow larger. Long ago it was thought that the world was only a few thousand years old. Today we believe that the earth is about five billion years old and that many parts of the universe are much older than that.

Scientists have made many observations about the universe that they cannot explain very well. For example, galaxies far from the earth seem to be moving away at very high speeds. If these observations are correct, all of the matter in the universe must have been lumped together some billions of years ago. Some scientists believe that a great explosion tore the universe apart about five billion years ago and started the galaxies moving away from each other. This is called the *big bang* theory of the universe.

Other scientists do not accept the *big bang* theory. They believe that the universe has been manufacturing new matter at a steady pace. This is called the *steady state* theory of the universe. Both theories explain some of the facts we have. Neither theory explains all the facts. So the chances are that both theories are somewhat wrong. This will give you an idea of the kinds of problems astronomers must try to solve.

Some of our difficulties in solving them come from the limited scope of our instruments. The whole universe is so large that there is probably much that we cannot see in the limited range of our telescopes. The earth's atmosphere also makes observation difficult. Many scientists hope that the new radio telescope will help solve some of our problems by *looking* deeper into outer space. Even if it does, there are more than enough problems to keep astronomers and physicists busy for many, many more years.

CHAPTER 3

THE SUN

If you look into the sky during daylight hours, you cannot see any stars except the sun. The glare of the sun's rays, dispersed through the atmosphere, makes it impossible for us to see the feebler light of the stars that are farther away. At night, when the sun is not shining on our part of the earth, we say the stars are *out*, forgetting that they have been there all the time. We sometimes forget that the sun is also a star. It is the closest star to earth, only 93 million miles away. We can see the sun very clearly; its light reaches us in about eight minutes. The star nearest us after the sun - named *Alpha Centauri* - is more than four light years from earth. That is about 25 trillion miles. So you can see that the sun is easier for us to study than any other star. Much of our knowledge of stars comes from studying the sun.

Facts About The Sun

The sun is a main sequence star about 860,000 miles in diameter. That is, its diameter is just over one hundred times the diameter of the earth. But this makes a much greater difference in volume. Both the earth and the sun are practically spherical. To find the volume of a sphere, we use the formula $V = 4\pi r^3/3$. Since $\pi = 3.1416$, to get just an approximate idea we can cancel out the π of the numerator with the 3 of the denominator. This gives us $V = 4r^3$. (That is not accurate, of course, but it gives a pretty close relationship.) So, Sun $V = 4 \times 430{,}000 \times 430{,}000 \times 430{,}000$; Earth $V = 4 \times 4{,}000 \times 4{,}000 \times 4{,}000$.

This shows that the sun has, roughly speaking, a million times as much volume as the earth. That is why we do not have to worry about the sun's running out of hydrogen fuel. It is so large that it can change vast amounts of matter into energy every second without a noticeable loss of mass.

The sun is all in a gaseous state. Temperatures on the sun are so high that we know it could not be in a solid state. We know that it has no solid crust, for parts of it turn at different rates of speed. If the continent of Africa, let us say, turned faster than the continent of North America, the earth's crust would break. On the sun, some parts do turn faster than others. The inside of the sun is probably made of hydrogen gas under great pressure. This hydrogen is consumed in the sun's nuclear furnace to produce energy and helium.

The Surface of the Sun

So the sun actually does not have a real surface at all. The face of the sun is a mass of gases hot enough to melt any metal we know. Iron and copper and the other elements found on the sun are all in a gaseous state because of the high temperatures.

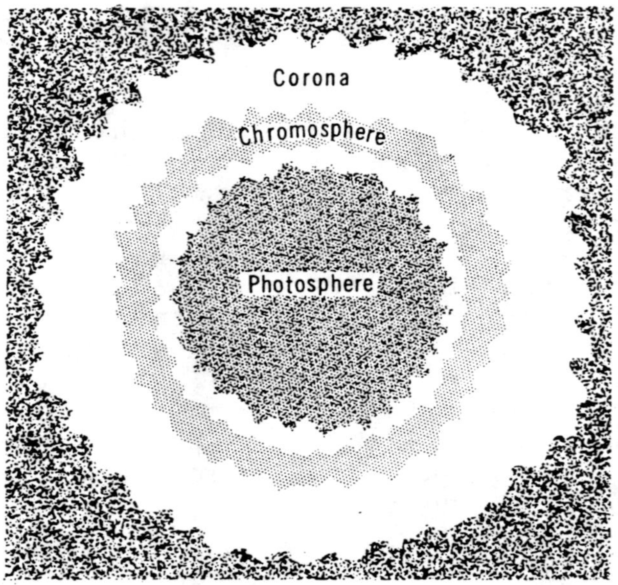

The disk of the sun is called the *photosphere*. Above the photosphere is a layer of flaming gas from five thousand to ten thousand miles deep. This layer is bright red in color. It is called the *chromosphere*. It could be compared to the atmosphere of the earth because it surrounds the sun's disk. Outside it, there is yet another layer of gases called the *Corona*, which extends perhaps a million or more miles into space around the sun.

The temperature around the sun changes. Close to the surface, in the photosphere, it is about 6,000°C. Higher, in the Corona, the temperature rises, reaching as high as 1,000,000°C at times. Even this is much lower, however, than the temperature at the center of the sun, which is estimated at 20,000,000°C.

Solar Flares

The sun as we see it seems to be a smooth, flat disk. But pictures reveal it as a swirling mass of flaming gases. The surface is like the inside of a furnace. Flames leap out, and great columns of fire, called *solar flares*, jump thousands of miles into space. The figure below shows a large solar flare in action. The white dot shows the size of the earth. This explosion threw a sheet of flame 250,000 miles into space. When the sun *kicks up its heels* this way, even the earth feels the shock 93 million miles away. Great clouds of electrical particles rain down toward the earth. They rush into our atmosphere and cause electrical static and other interference with radio messages.

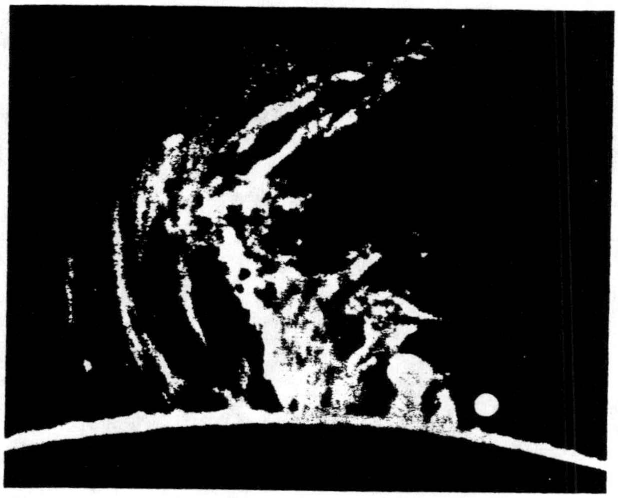

Inside the Sun

We can see only the surface of the sun. But by studying the surface we can learn something about the inside of the sun too. We now know that nuclear reactions are always taking place inside the sun, just as they occur inside a hydrogen bomb. This was not known until World War II. While looking for new military weapons, scientists found that the energy inside the atom could be released from certain elements. They learned how to split the nucleus of the atom. This, as you know, is called nuclear fission. Later, after World War II had ended, scientists also learned how to put together atoms to release even more nuclear energy by nuclear fusion. Nuclear fusion is the principle used in the hydrogen bomb. The sun, also, carries on nuclear fusion inside its surface. The temperature is very high, and nuclear fusion goes on continuously. The energy thus released is thrown off as radiation.

Every second the sun converts over *four million tons* of matter into energy. It has been using up this much of its matter every second for millions of years. Yet the sun is so large that only one millionth of one millionth of its mass is consumed in an entire year. The sun has an ample supply of fuel for a long time.

The Sun and the Earth

The sun is the most important star in the whole universe for us living creatures on earth. We would all die if the sun stopped radiating. Except for the small bit of atomic energy we produce and that from deep inside the earth, every bit of energy on earth comes from the sun. Green plants are able to use the energy from the sun to make chemical changes inside their own bodies. They trap the sun's energy and change it into plant material. Animals eat the plants and change plant material into animal material. Even the energy in coal and petroleum was once part of the sun. Plants that had stored solar energy in their bodies were changed - under great pressure through the ages - into coal. When we burn coal, we simply release energy that some plant took from the sunlight long ago. Only in an atomic reactor or through chemical reactions can energy be produced that does not originate in the sun.

A very small fraction of the sun's total radiation reaches us on earth. If much of it did, we would all be burned to a crisp. Only a few billionths of the sun's output falls on the earth's surface. This tiny fraction supplies the energy that warms our planet, moves the winds and the oceans, and grows our whole food supply. Without the energy from sunlight, no plant would grow, no rain would fall, and all of our water would quickly turn to ice.

Every minute millions of tons of water are lifted into the air by solar energy. This water is carried over the earth by winds, which are generated by solar energy. Thus, the sun's heat is distributed evenly over the earth. The sun is like the central unit in a giant air-conditioning system. It provides the heat that keeps us warm. It also runs the carrying system that circulates the heat. From no other source could we begin to replace the energy which the sun provides for us. All of the energy produced by a million atomic bombs would not equal the energy we get from the sun in just a few days.

Sunspots

Although we know a great deal about our own private star, the sun, there are still some things about it which remain mysteries to us. One of the most interesting of these mysteries is the occurrence of great whirls that appear on the surface of the sun from time to time. These whirls look like dark spots on the face of the sun. We call them *sunspots*. Close-up pictures show them as great whirling masses of gas that move across the sun's disk. There are usually two or three of them on the surface at the same time. They may last only a few minutes, or they may stay several weeks. Some are small - only a few miles in diameter. Others are larger than the earth. We know they are not *hotspots*, for the temperature is actually cooler inside the sunspot than on the rest of the sun's surface. We do not know why they appear. They seem to come in regular cycles. Every eleven years the number of sunspots increases rapidly. For a short time there are a great many large ones. Then the activity slows down again.

That is about all that we know about sunspots at this time. They have been watched and charted. Photographs of them have been taken. But why they appear or what they are remains a mystery. We do know that their effect is felt even here on earth. They seem to cause tremendous electrical storms in the atmosphere far above the earth. Radio signals often *fade out* at this time. Sometimes intercontinental radio is interrupted for several days. The *great northern lights*, *aurora borealis*, becomes active. These glowing streaks of electrical energy appear around the poles of the earth like giant streamers. Even on the surface of the earth, snowstorms and electric storms are more violent during a sunspot period than during normal times. Here is one of nature's mysteries, right on our own doorstep, that we have not been able to solve. There are also other questions about the sun that we have not yet been able to answer.

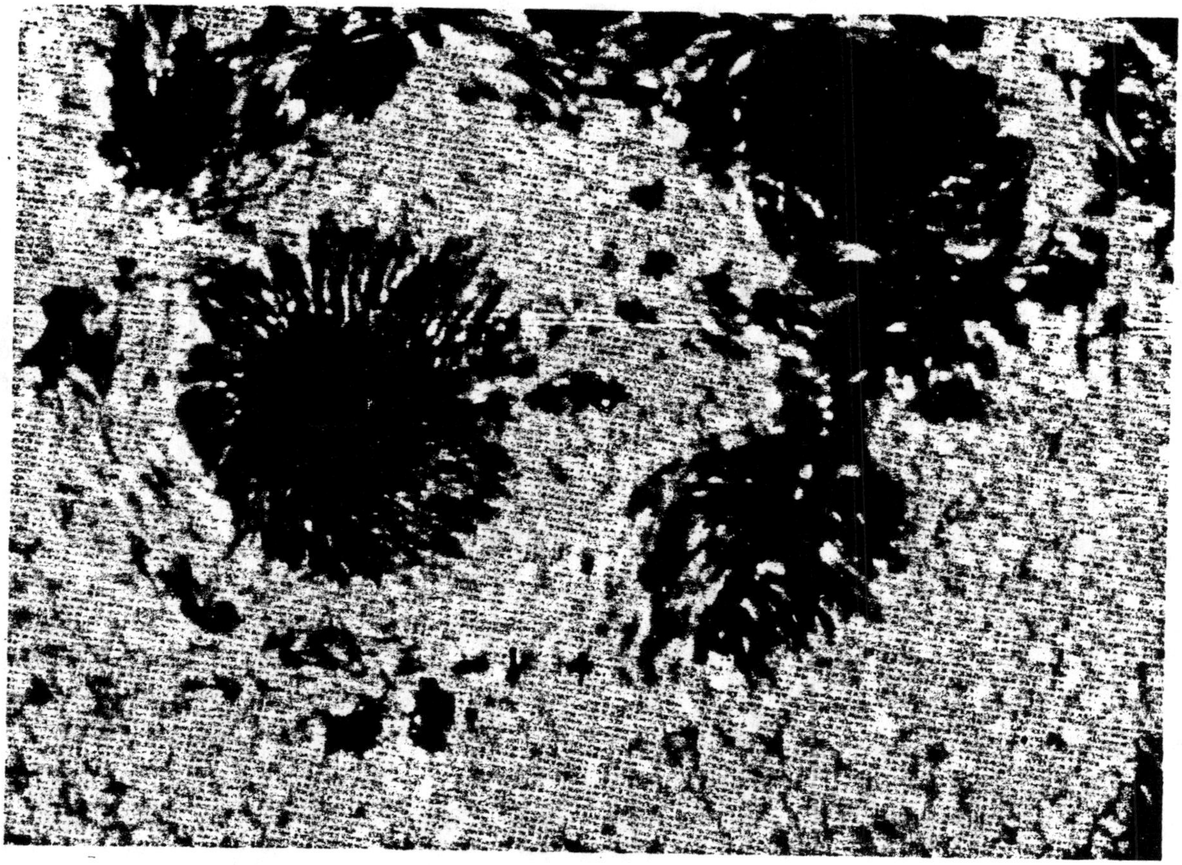

Using the Sun's Energy

When we buy gasoline, or coal, or food, we are actually buying energy to run our automobile, or heat our house, or keep our bodies alive. Such commodities are expensive, yet we must have energy to live. Every day the sun pours great amounts of energy onto the earth. Sunlight is free for everyone. Why cannot solar energy be used directly to do some of the jobs we use coal and oil for now? The answer is that solar energy can be so used, but it is often cheaper to produce energy by burning coal than to harness the energy of the sun. Nevertheless it is possible to use solar energy to cook, to heat houses, and even to make electricity.

If, on a hot summer day, you leave your car for awhile with the windows closed, it becomes very hot inside. The car acts as a *heat trap*. Sunlight passes freely through the windows; it strikes the car seats and other parts of the interior; and some of the energy is absorbed, or taken up, by the material and changed into heat. The rest is reflected back out the windows. But a good deal of the heat inside the car cannot escape through the windows. Glass, which lets light in freely, is not a good conductor of heat. It is an insulator. Thus, much of the heat is held inside the car, and the temperature rises steadily as the sunlight pours in.

If there is enough sunlight, this same principle can be used to heat a house. The top of the house can be covered with glass panes. Beneath the glass, the roof can be painted a dark color so it will absorb sunlight well. The sunlight passes through the glass and is absorbed by the black paint. The heat cannot escape through the glass. So the *heat trap* becomes very hot. Water can be circulated through this *heat trap*. The heat warms the water, which can then be circulated through the house to warm it. Actually water has been known to boil in a *heat trap* of this sort. But if the skies are cloudy and there is not enough sunlight, this system will not work well. It is used now in some homes in the southern United States, where there are many sunny days.

Another method of producing heat from sunlight is based upon the same principle as the reflecting telescope. A large reflector is used to concentrate the rays of the sun in a small area. When the sun's rays strike a pan placed in the center of the reflector, they are changed from radiant energy to heat energy. A reflector a few feet wide can boil water very quickly. By using reflectors more than a hundred feet in diameter, scientists can produce very high temperatures with the rays of the sun. Temperatures of several thousand degrees C can be produced if the rays are captured from a large area and concentrated into a small area.

Notice that sunlight is not heat. Sunlight is radiant energy. It can be changed into heat, which is another form of energy, if it is absorbed by the surface it strikes against. When we wear dark-colored clothing in the summertime, the cloth absorbs energy from the sun and changes it to heat. This is why people who live in warm climates wear light-colored clothing. Light-colored clothing reflects the sun's rays instead of absorbing them.

CHAPTER 4

THE SOLAR SYSTEM

The sun is the center of a system that contains many different bodies which orbit around it. The sun together with all its satellites is called the *solar system*. The largest of these satellites are the nine *planets*. The earth is one of these nine planets. As the earth circles the sun, it takes with it a satellite of its own, the moon. Several of the other planets have moons like ours also. These moons are all parts of the solar system too. So are the thousands of smaller bodies, called *meteors* and *comets*, that orbit around the sun.

Most of the matter in the solar system - about 98 percent, in fact - is concentrated in the sun itself. Only 2 percent is found in all the rest of the bodies together. That is, the sun is fifty times as massive as all the rest of the solar system.

The Origin of the Solar System

Where did the solar system come from? How did these celestial bodies originate? These are questions that scientists have not been able to answer as yet. Scientists, called cosmologists, have been studying the question of the origin of the solar system for a long time. They have come up with several theories to explain it, but as yet they have not found one that all the experts in the field can accept. As new facts come to light, any theory has to be adjusted to contain them. If they will not fit, the theory has to be discarded. This has happened to several theories about the origin of the solar system. Cosmology offers a very good example of the way theoretical scientists work. They must constantly revise their ideas to fit new knowledge as it is revealed.

Three theories of the origin of the solar system seem to be supported by more scientists than others. They are worth a brief explanation. It will give you an opportunity to see how scientists form their hypotheses. One is that long ago a giant star passed near the sun. The force of gravity drew a great mass of gaseous matter out of the sun. As the star passed by, this mass of matter escaped from union with the sun and formed, instead, a series of rings around it. Then the escaped gases condensed into liquid and finally into the solid planets and other bodies that we know in the solar system.

The second theory proposes that there were once two suns. Since most stars seem to live in pairs, this is a reasonable guess. The theory then assumes that the second star exploded. After the explosion, the earth and other planets were formed from the matter that remained. A variation on this theory is that the second star never became formed actually. Instead, the gas from the second star was distributed through space and was formed into the planets and moons.

Another theory, once very popularly believed, is that the solar system was once a mass of dust and gas. The gases condensed, and the star that we call the sun formed in the center. The material that remained outside the central body formed the planets and other members of the solar system.

The Present Solar System

Although we do not know how the solar system began, we do know what it looks like at the present time. There is the sun in the center of the whole system. The sun is one star, held in place in a very large galaxy called the Milky Way. Around the sun, there are the nine planets and thirty-one moons, as well as thousands of small planets called *asteroids*. There are, also, millions of chunks of matter called *meteors*. And there are millions of clusters of frozen particles called *comets*.

You can see the pattern of the solar system in the figure below. We are looking down at the top of the system in the drawing. Each planet has an orbit, or path around the sun, that it follows regularly. The sun alone radiates heat in the solar system. All other members of the family are dead bodies that obtain their heat from the sun.

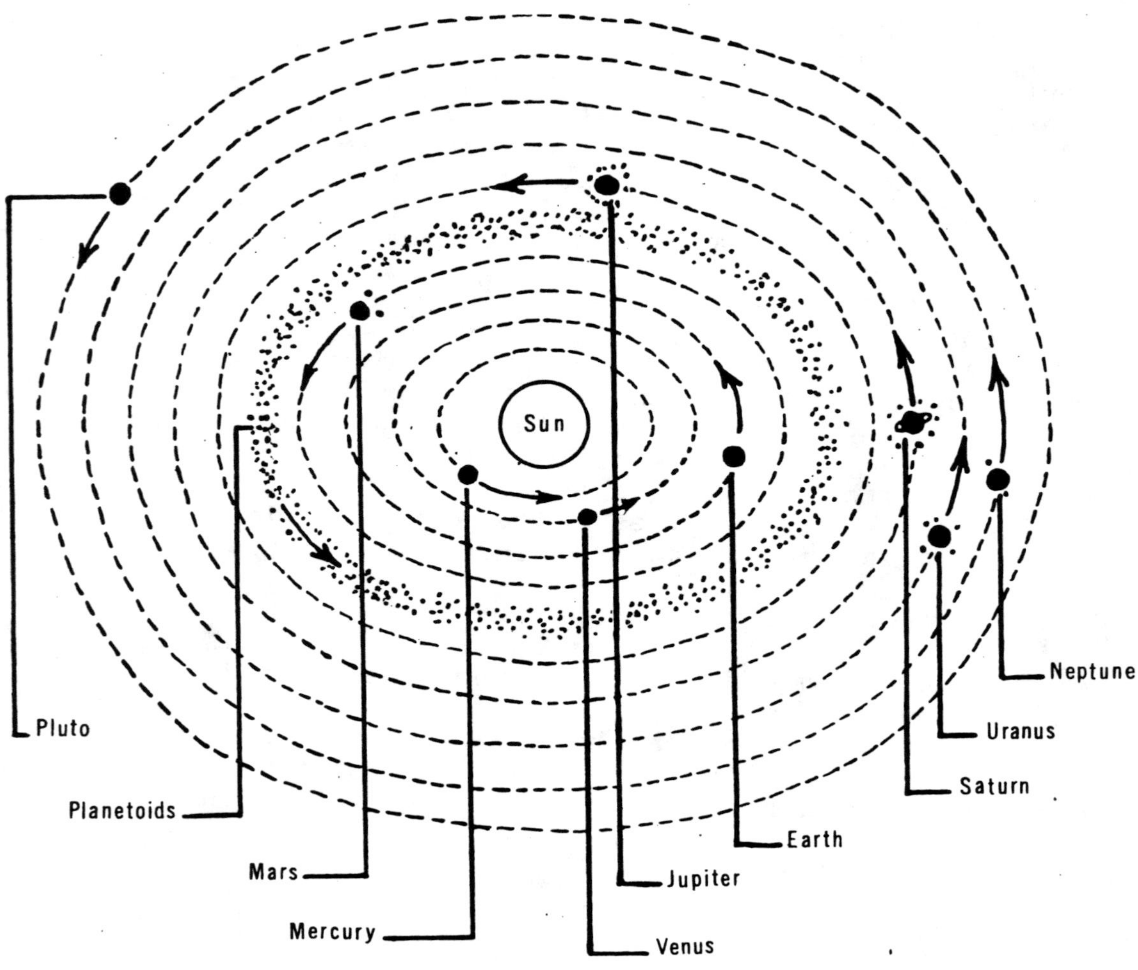

The Planets in the Solar System

We can see the planets because they reflect light from the sun. Five of the planets (Mercury, Venus, Mars, Jupiter, and Saturn) can be seen with the naked eye. The others must be viewed through a telescope and even then they are not easily spotted. The one that is farthest from the sun, Pluto, has been photographed through a large telescope. But little is known about it.

The sun is not in the exact center of the orbits of the planets. For example, the earth in one part of its orbit comes within 91 million miles of the sun; but in another part the two are nearly 94 million miles apart. The orbits of the other planets, too, are not quite circular. They are elliptical. That means they are like circles stretched out.

The Size of the Planets

The chart that follows shows the diameter of each planet and the distance from the planet to the sun. Notice that there are two groups of planets. Mercury, Venus, Earth, Mars, and Pluto are much smaller than the four large planets Jupiter, Saturn, Uranus, and Neptune. The four large planets are called *Jovian* planets. They are four to eight times as large as the earth in diameter. The smaller planets are called *terrestrial* planets. *Terrestrial* means *like the earth*. These smaller planets are all about the same size as the earth.

A Closer Look at the Planets

Since we live on a planet, it is interesting to see what the other planets in our solar system are like. Actually, none of them are much like our own. Mars is probably the most like Earth of all other planets. Each one is different from the others in climate, appearance, size, and many other features.

Chart of Planets

Name	Diameter (miles)	Distance from the Sun (million miles)
Mercury	3,100	36
Venus	7,900	67
Earth	8,000	93
Mars	4,200	142
Jupiter	88,600	483
Saturn	74,100	886
Uranus	32,000	1,782
Neptune	31,000	2,793
Pluto	(not known)	4,000 approx.

Mercury, which is closest to the sun, is the smallest, coldest, hottest, and fastest planet in the solar system. It moves around the sun at a speed of about twenty-three miles each second. Every eighty-eight days it completes one rotation and one revolution. Thus its day and year are of equal length. Its movements are such that the same side is pointed toward the sun at all times. Because Mercury is only 36 million miles from the sun, the side facing the sun is very hot. The side away from the sun is very cold. The temperature averages about 400°C on the hot side, and about -200°C (200° below zero) on the cold side. You can see that life as we know it cannot possibly be found on Mercury.

The largest planet in the solar system is Jupiter. This large body spins very rapidly on its axis. Every ten hours Jupiter makes one complete revolution. It has a day ten hours long instead of twenty-four as that of Earth. The distance around the equator of Jupiter is almost eleven times that of Earth.

Venus and Saturn are very beautiful planets. Saturn has broad white rings around its equator. These rings, two bright ones and a fainter inner one, are tiny particles caught in the gravity around this planet. The particles orbit around Saturn like a well-disciplined parade. This produces the band effect you can see in the figure below. Venus is one of the brightest objects in the night sky. This planet is covered with a solid mass of white clouds, which reflect light very well. Venus is warmer than Earth, but its temperature is not so high as that of Mercury.

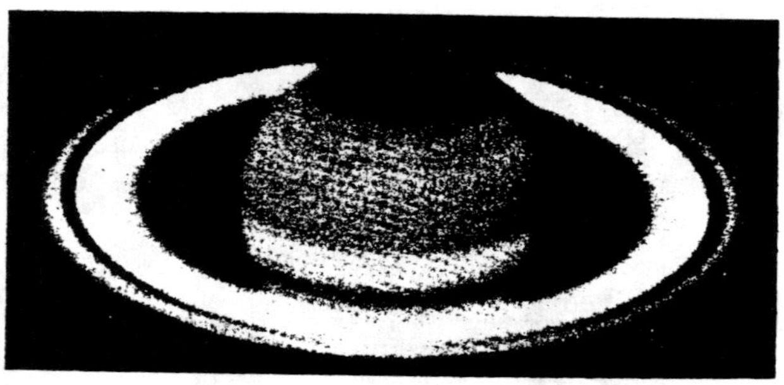

Pluto, the loneliest planet, was not discovered until 1930. We know that Pluto is about the same size as Mercury and that it is very cold because it is so far from the sun.

Mars

Since we live at the beginning of the space age, the planet Mars is most interesting to us. Mars is the first planet that man has visited. Mars is more like the earth than any other planet in the solar system. It has clouds, and it has some atmosphere. There is probably some water on Mars too, since the poles appear to be white during the cold season, suggesting that snow and ice form there just as at the North and South Poles of the Earth. Most of Mars seems to be a dry desert area. Parts of Mars change color during the year. This may mean that there is plant life of some sort on the planet. Some people believe that animal life may exist on Mars too, but this seems less likely.

In 1877 a number of long straight lines were discovered on Mars. These are the famous *Schiaparelli's Canals*, named for the Italian astronomer who discovered them. The lines are so long and straight that people believed that they had been built by some animal. But astronomers think that the *canals* are not manmade at all. Clouds obscure our view of Mars, and no one is certain what the dark lines really are. The problem could be solved easily by landing on Mars or even by taking a photograph from a close distance. This was accomplished in the late 90's with our unmanned landing on Mars.

Moons

We are all familiar with our own natural satellite, the moon. We will study it in detail in the next chapter. You may not know that there are thirty other moons in the solar system, some larger than ours. Mars and Neptune have two moons each. Jupiter has twelve. Saturn has nine, and Uranus five. Six of the moons are larger than ours. In fact, four of Jupiter's moons are nearly as large as the planet Mercury. These moons are not very important in the solar system. Our moon is chiefly responsible for the tides, and it does reflect some light to the earth at night. We do not know what effect the other moons in the solar system may have on their planets.

Since 1957 the earth has been presented with several new *moons*, or artificial satellites. Some of these satellites are in orbit for just a short time and then plunge back to earth. Others will remain in orbit almost indefinitely. These artificial satellites also help us learn more about the universe. Already they are used for weather prediction, navigation, communication, and many other uses. They have also provided us with information about radiation in the space beyond our atmosphere.

Comets and Meteors

A comet is a chunk of matter made up of frozen particles and gases. The main body of a comet is usually about one mile in diameter. We think of a comet as an object with a long *tail*. But this tail is present only part of the time. While the comet is far out in space, it has no tail. When it approaches the sun, the heat melts the frozen material and releases gases. These form the comet's long tail we see so often. This tail may be thousands of miles long.

One very famous comet, called Halley's Comet, was first sighted long before Christ was born. Halley's Comet was last in our area in 1986. Its orbit brings it back to the sun once every seventy-six years. The comet is visible for several days as it orbits near the sun. Then it heads back into space, losing its long white tail and freezing solid again.

Any piece of solid matter flying through space is called a meteor. There are millions of meteors in the solar system alone. Most of them are about the size of a pea. A few are large and heavy and weigh many tons. Astronauts will have to worry about these meteors when they leave the earth's protective atmosphere. Every day thousands of meteors strike the earth. These fallen meteors are called *meteorites*. When they enter the atmosphere, friction between the meteors and the air molecules produces heat that burns all but a few of the meteors. Those that strike the earth are usually made of iron or rock. Some of these meteorites are quite large. The Barrington Crater in Arizona was made by a meteorite. This crater is one-half mile wide and about five hundred feet deep. Another large meteorite landed in northern Russia. The noise was heard for hundreds of miles. Trees were knocked flat for miles around the crater. The meteor dug a great hole in the earth and buried itself.

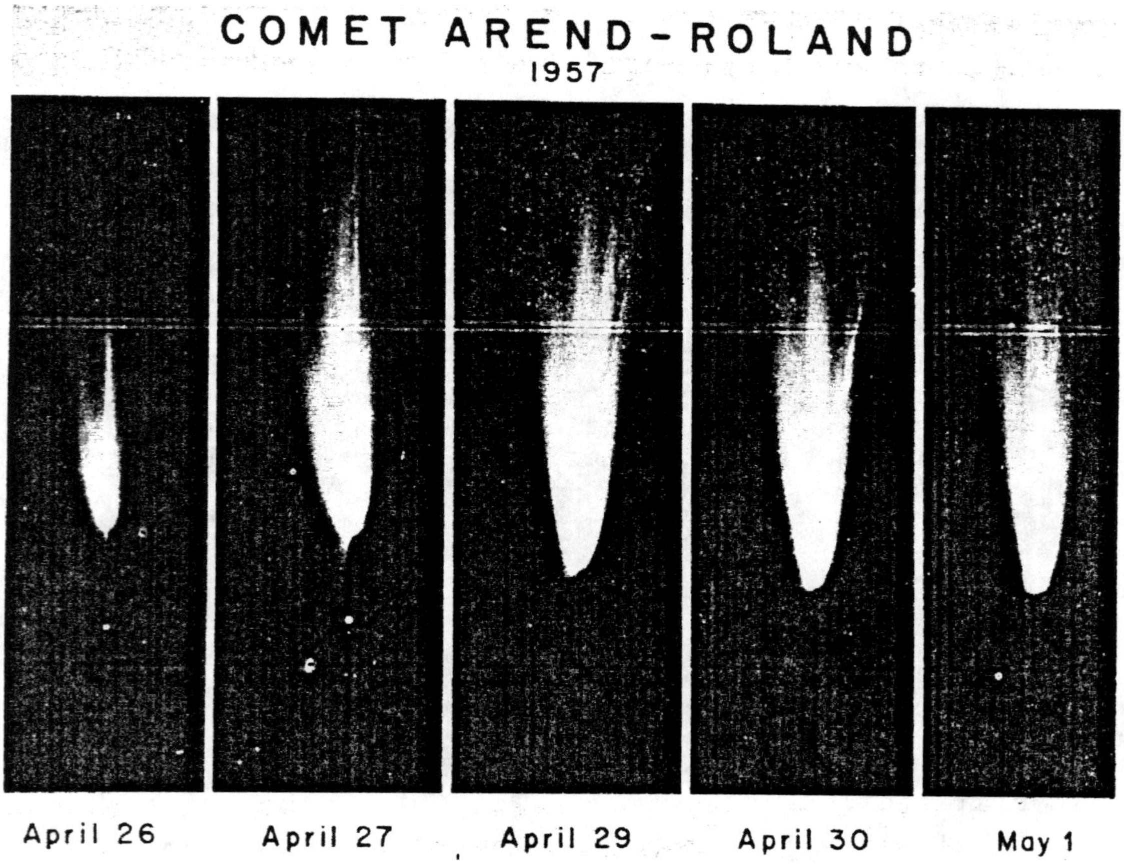

CHAPTER 5

THE MOON

The earth's one natural satellite we call the moon. This satellite orbits around the earth just as the earth orbits around the sun. Many scientists believe that the moon was once part of the earth. They believe that ages ago it was torn from the part of the earth where the Pacific Ocean basin now is.

But a good many scientists do not accept this theory. They believe that the moon was formed separately. Neither theory can be proved at this time.

Size and Location of the Moon

The moon is a sphere, or round ball, about two thousand miles in diameter. It is, roughly speaking, 1/60 as large as the earth. The moon's orbit is about 240,000 miles from earth. It makes one complete circuit in 27 1/3 days. Like the earth, the moon also rotates on its axis as it travels its orbit. It, too, makes one full rotation every twenty-four hours. Since both the moon and the earth rotate at the same rate, the same face is always directed toward the earth.

The moon seems to be made of the same elements we find on earth. That is why some people believe that the moon was once part of the earth. Since the moon is the smaller, it does not have as much attraction for other bodies as the earth does. The mass of the moon is less than that of the earth. For this reason your weight, if you were on the moon, would actually be only 1/6 of your weight on earth. Since your muscles would be the same, you could probably break the world high jump record just hopping up and down.

Man landed on the moon in 1969 and returned several times.

Phases of the Moon

In the night we can see the moon shining in the sky. But we know that the moon is a dead body like the earth. It does not radiate energy. It shines by the reflected light of the sun. Since the moon moves around the earth and the earth moves around the sun, we cannot always see the moon. In parts of the orbit, the moon comes between us and the sun. When it does, only a small part of the moon can be seen from the earth. These changes in the amount of light reflected from the moon are called the *phases of the moon*. The part of the moon we can see depends upon the relative position of the moon, the earth, and the sun.

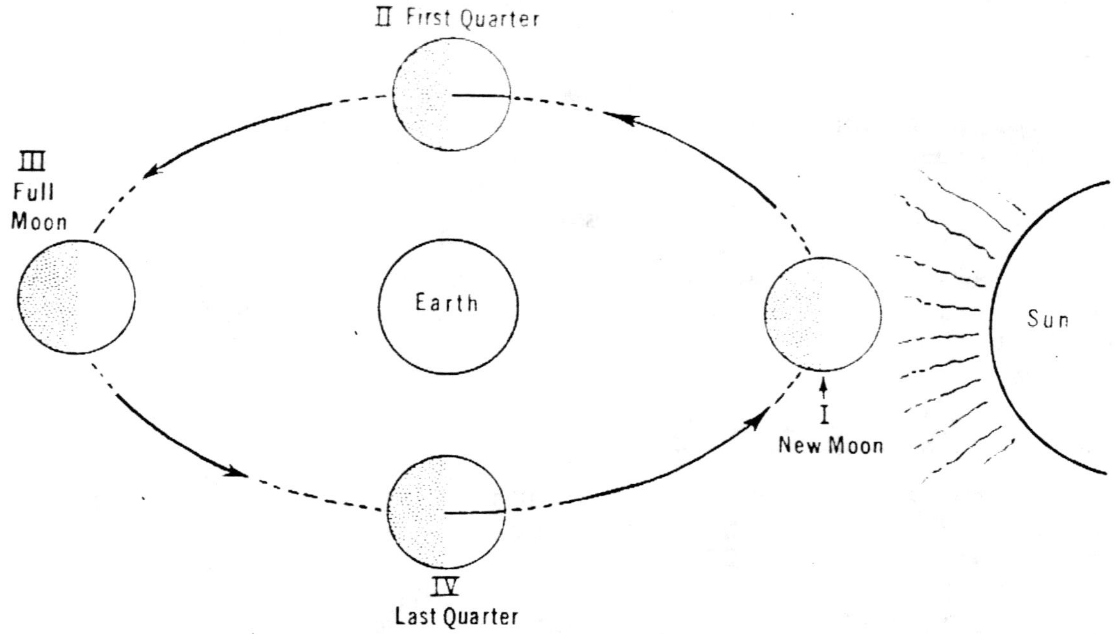

In step I of the figure above, the moon is between the sun and the earth. The side of the moon facing the sun is lighted. But this side of the moon cannot be seen from the earth. Therefore, the moon appears to be dark because the side toward the earth is not lighted by the sun. This is called the *new moon*.

The moon continues to orbit around the earth, moving counter-clockwise. That is, it moves in the opposite direction to the hands on a clock. When the moon reaches step II, 1/4 of its lighted surface can be seen from the earth. This is called the *first quarter* of the moon.

In step III, the earth is between the moon and the sun. Now we can see the full face of the moon. This is called *full moon*. It is the exact opposite of a new moon. After full moon, the moon gradually appears to grow smaller and smaller until it disappears once again in a new moon period. Of course, the moon does not stop at each step in the drawing. It passes through these four positions as it moves around the earth. As it looks to us, the moon seems to grow from a thin sliver to a full moon and then to fade away to a thin sliver again. You can see the face of the moon in the figure below as it appears during the full moon and two quarters.

The Surface of the Moon

Look closely at the surface of the moon in the figure above. Notice the great pits and holes in the surface. These scars are probably due to meteors striking the moon. There is no blanket of atmosphere on the moon, and so the surface gets the full force of any meteor that hits it. Another possible cause of these pits could be volcanic activity. Some astronomers believe that at some time the moon may have had some great explosions that tore up its surface. One reason why the craters are hard to explain is that they are so large. We do not have pits and craters of this size on earth.

The mountains on the moon have peaks as high as those here on earth. Some of them are more than 25,000 feet high. Because there is no air or water on the moon, there is no wind and no rain. You could probably feel sound vibrations on the moon with your feet, but there would be no vibrations against your eardrums because there is no air to carry the sound. So there would be no sound except that carried through the solid parts of the moon. Since there is no weather on the moon, there is no erosion except for that caused by falling meteorites. The surface of the earth is worn away by the action of wind and water and glaciers. On the moon, the surface remains almost unchanged. The mountain peaks are high and sharp. They have not been worn down like our mountains on earth.

The surface of the moon was thought to be covered with a thick layer of dry powder or dust. The surface is solid rock with many craters caused by meteorite impact. Some were as little as three feet across.

Temperature on the Moon

A weatherman on the moon would have an easy job. Every day is clear, sunny, and hot. Every night is dark and very cold. The side of the moon facing the sun is hot. In fact, the temperature is 100°C, hot enough to boil water. The dark side of the moon is cold. Usually the temperature is -130°C, or 130° below zero. The temperature on the moon drops very quickly when the sun goes down. In a few hours the temperature might drop nearly 200° on the Centrigrade scale. This drop has been measured during an eclipse. As the earth moves between the sun and the moon and cuts off the sunlight, the temperature in the shaded area of the moon falls rapidly.

Eclipses

When the earth moves between the sun and the moon, it casts its shadow on the moon. Light travels in a straight line and it cannot bend around the earth to light the moon. When the moon moves between the earth and the sun, the moon's shadow falls on the earth. Within the shadow, the sun cannot be seen. Since the moon is smaller than the earth, its shadow covers only a small part of the earth's surface. Outside this shadowed area, the sun is bright and clear. Inside the shadow, the sun is blotted out by the moon. This shadow period is called an *eclipse*. When the earth blots out the moon, we have a *lunar eclipse*. When the moon blots out the sun, we have a *solar eclipse*. A lunar eclipse occurs about every six months. Several years pass between total solar eclipses.

A Lunar Eclipse

Every month the earth moves between the sun and the moon. But it does not always cause an eclipse. Sometimes the earth's shadow falls above or below the moon and there is no eclipse. When the earth's shadow does strike the moon, we have a lunar eclipse. A lunar eclipse is shown in the figure below. The earth, sun, and moon are all in a line. The shadow of the earth covers the moon. No light from the sun strikes the moon and, therefore, no light can be reflected. We cannot see the moon at all in lunar eclipses.

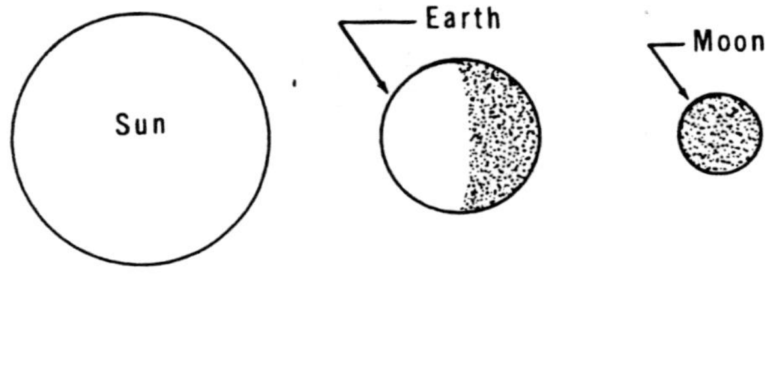

A Solar Eclipse

When the moon comes between the earth and sun, we have a solar eclipse. In a small area on earth, perhaps a hundred miles square, the sun is then completely blotted out. The position of the three bodies is shown in the figure below. A total eclipse lasts only a few minutes. The location from which the total eclipse may be seen changes. A total solar eclipse was visible from the state of Maine in 1963 and another was visible from Florida in 1970. After 1980, there will be no more solar eclipses visible from the United States until the twenty-first century. The location of an eclipse is known many years in advance. We also can know of those long past, because the motion of the solar system is regular so that we can calculate when they occurred.

Solar eclipses have been very useful to science. As the sun is blotted out by the moon, scientists can study the sun's surface more carefully. Einstein's theory of relativity was first tested during an eclipse. He predicted that the mass of the moon would bend the light rays from the sun. His theory was proved to be correct. The theory plays a vital part in the scientific study of the universe.

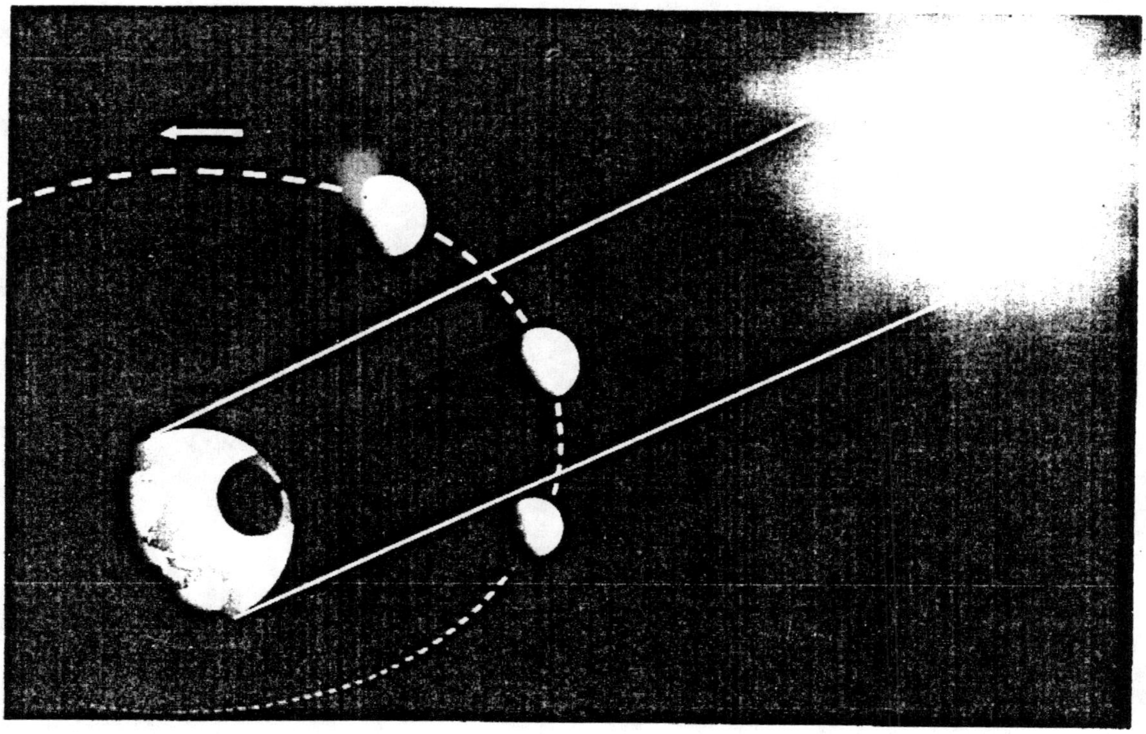

Tides

Anyone who lives in an ocean harbor knows what an important part the tides play in ocean shipping. The chief cause of tides is the moon. The attraction of the moon for the earth causes the water on earth to rise toward the moon. This creates a bulge in the oceans that follows the moon around the earth. When the bulge reaches shore, we have tides. During a tide the water level may rise several feet and then fall back once again as the moon passes. Some harbors in the world can be used only when the water level is at high tide. By their rise and fall the tides help clean out the bottom of the harbors.

Although the sun and moon are the main cause of tides, the shape of the shoreline and the depth of the ocean also affect tides. Some parts of the world have little or no tides. Other areas have tides that raise the water level twenty feet or more each day.

CHAPTER 6

THE EARTH

Suppose an astronomer outside the solar system could tell us how the earth looks to him. What would he say? First, he would tell us that the earth is a dead body and not a star. The earth is a planet of the sun, moving around it in an orbit that is nearly a perfect circle. The sun is not exactly in the center of the circle, and the earth is closer to the sun in January than in July. Our astronomer would say that the earth is just over eight light-minutes, or 93 million miles, away from the sun. He would then go on to describe the clouds he sees and the water on the earth's surface. He would probably not mention people, since it would be hard to see anything this small. An elephant standing on the moon could not be seen from the earth even with a powerful telescope.

The Planet Earth

The earth is like the other planets in the solar system in some ways. It is almost spherical in shape. It moves counterclockwise around the sun, as the other planets do. It also depends upon the sun for its energy supply. The earth spins on its axis, but not at the same rate of speed as the other planets. The earth is not quite spherical, as it is flattened slightly at the top and at the bottom. The length of a line drawn around the earth and through the poles is about sixty miles less than one drawn around the Equator. The difference is due to the polar flattening at the top and at the bottom of the earth. Three-fourths of the earth is covered with water. The rest is dry land. But in many areas only a small part of the dry land is actually used.

The earth differs from the other planets in some important ways. It has an atmosphere that contains free oxygen. The atmosphere, as well as the surface of the earth, contains a great deal of water. Millions of plants and animals live on the earth and in the oceans. Unlike most of the other planets, the earth has climates and temperatures that support plant and animal life.

Movement of the Earth

Even when we stand perfectly still, we are moving through space in three directions at once. As part of the earth, we move with the surface as the earth spins. As part of the solar system, we also move in an orbit around the sun. And, finally, as part of the Milky Way galaxy, we follow the sun as it sweeps along in its place in the galaxy. We take part in all of these movements whether we like them or not. We are so accustomed to our environment that we do not even notice the motion. The two movements of the earth that concern us most are the revolution of the earth around the sun and the rotation of the earth on its axis. These two movements give us day and night, the seasons, our calendar and time system, and a measure of distance.

Measuring Time

As you have learned, the earth makes one complete turn on its axis every twenty-four hours. This gives us our standard for time. The day is divided into twenty-four equal hours, which contain altogether 86,400 equal parts called seconds. The spinning of the earth thus supplies our standard clock.

For years, we were unable to check the accuracy of our natural clock. But today we know that it is actually slowing down. The earth does not turn quite as rapidly as it once turned. Every thousand years, it loses forty-five minutes of time. This slowdown is due to the force of the tides. When they work against the earth's rotation, they absorb some energy from the earth.

The earth makes one complete trip around the sun in one year. During the year, the earth turns on its axis 365¼ times. Our calendar is meant to keep track of days and years. But, since it contains only 365 full days in four years it loses the four quarter-turns of the earth, or one full day. So we add one day to the calendar every four years. The year with the extra day is called *leap year*. Actually, even leap year does not make the calendar completely accurate. The earth does not make exactly 365¼ turns in one year. So we really would need to make another even smaller correction to keep the calendar perfectly right.

Time Zones

If you roast a piece of meat over an open fire and use a spit to turn it, you know that each part of the roast passes over the fire as the spit is turned. The earth is like a roast on a spit. Each time it turns, it passes under the heat from the sun, and we are kept warm by the sun's heat. The earth makes one full turn every twenty-four hours. So each hour, it turns 1/24 of the total distance around the earth. Now the earth is a globe and the total distance around any globe is 360 degrees. So the earth turns 1/24 of 360 degrees in one hour. Thus, we can say that the surface moves fifteen degrees in one hour.

Look at the figure below. The globe has been divided into fifteen degree units. If the sun is exactly over the line marked A, it will be exactly over B in one hour, and exactly over C in two hours. The earth spins from west to east. So we know that a city at point A on the globe would see the sun overhead exactly one hour before a city on line B. And the city on line B would see the sun exactly one hour later than the city on line A

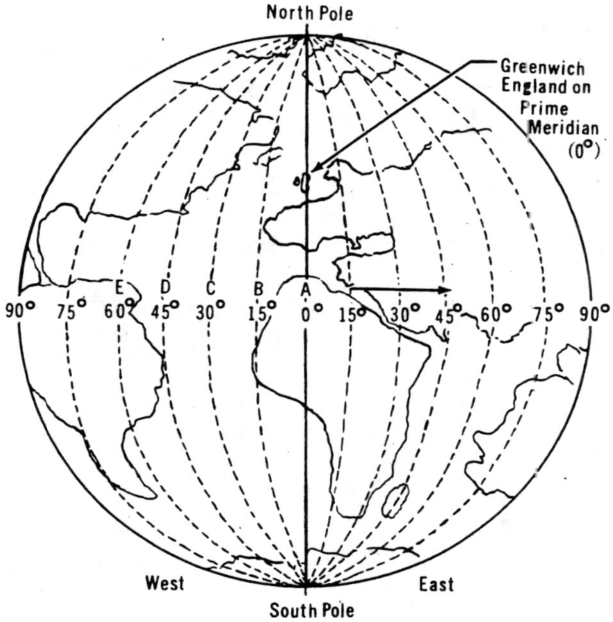

To keep all of the clocks on earth in time, we have divided the earth into twenty-four slices, each slice fifteen degrees wide. We have agreed that the time within each slice, or zone, will be considered exactly one hour different from the time in the next. Four of these time zones fall within the United States. The time zone on the east coast is exactly three hours ahead of the one on the west coast. The lines that divide the time zones do not fall exactly upon the meridians. They are drawn to avoid cutting through large cities and heavily populated areas.

The Prime Meridian

When the earth had been divided into the twenty-four time zones, we still had to decide at which zone the day was to begin. This was decided by international agreement. The day begins when the sun crosses the line drawn through the North and South Poles and the town of Greenwich, England. This line is called the *prime meridian*. The line is drawn completely around the globe. When the sun is exactly over the line on the opposite side of the earth from Greenwich, the day begins. When the sun is exactly over the line drawn through Greenwich, it is exactly noon at Greenwich. This is the standard for time throughout the world. Every time zone sets its watch by the Greenwich zone. The New York City time zone is always five hours behind the Greenwich zone. The Los Angeles time zone is eight hours behind Greenwich time.

The International Date Line

With twenty-four time zones and a prime meridian, we have solved most of our problems. But one time problem still remains. Suppose a pilot leaves Chicago in a jet plane. He flies westward toward the sun and chases it all around the earth. When he lands in Chicago, he has been flying in daylight throughout the trip. No night has fallen, and he may think he is coming back to Chicago on the same day he left. This would not be true, since night would have fallen in Chicago before he returned. The question is: when did he pass into the next day?

This problem is solved by drawing a line through the Pacific Ocean on the globe. We call this the *international date line*. Any traveler crossing this line from east to west must add one day to his calendar. Any traveler crossing the line from west to east must subtract one day from his calendar. This solves the problem for our jet pilot. When he crosses the line, he adds one day to his calendar. When he returns to Chicago, his calendar tells him that one full day has passed since his departure.

You will notice that the date line does not follow the meridian exactly. It is drawn, as all other zone lines, to avoid cutting through thickly populated areas.

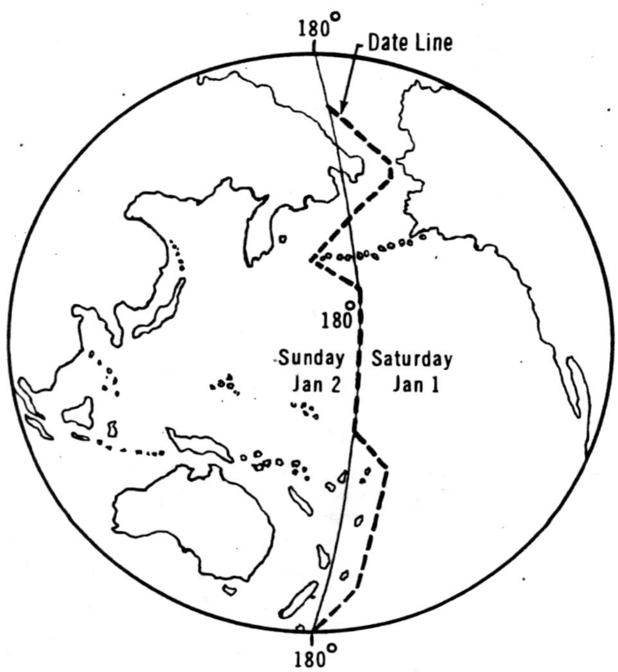

The Seasons

In the Temperate Zone where the United States is located, each year brings four seasons. In the northern United States, the difference between these seasons is striking. Winter is followed by spring, then summer, and then fall. The climate changes from cold to warmer, to hot and then to cooler as the change in season takes place. Why do these changes take place? There are two connected reasons. First, the earth moves around the sun in its orbit once each year. But this alone will not make the difference. Nor does the distance between the earth and the sun matter. The reason the seasons change is that the earth is not in an upright position. It is tilted to one side. The earth remains tilted in the same direction as it moves around the sun. It is this tilting, together with the movement around the sun, that gives us a change of season.

Look at the figure below. Notice how the earth is tipped to the right. It is 23½ degrees from a vertical, or up and down, position. On June 22, the earth is tilted toward the sun. The North Pole is exposed to the sunlight and the South Pole is hidden. The northern half of the earth receives more sun during this period than the southern half. So it is summer in the Northern Hemisphere.

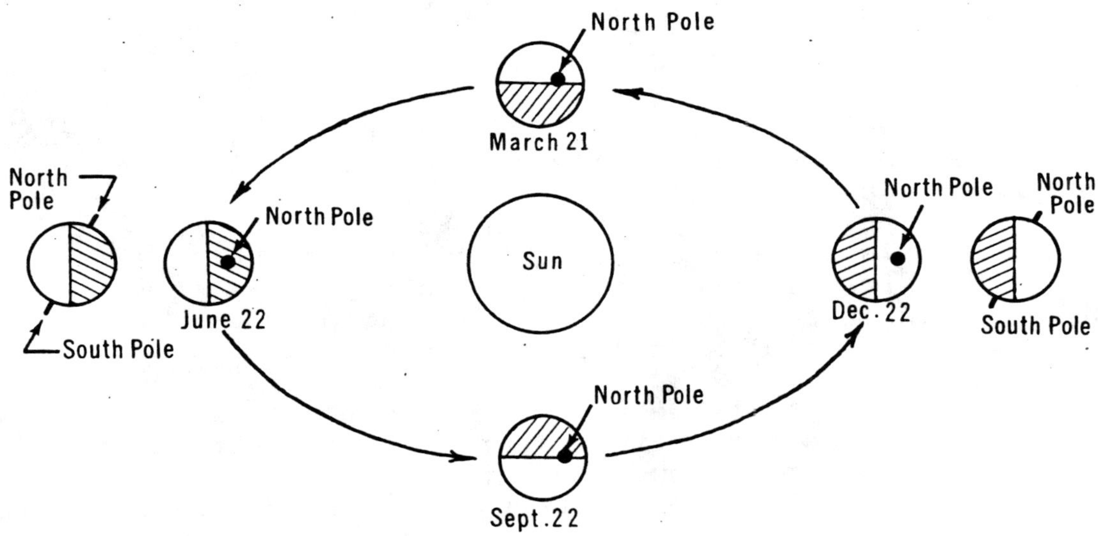

From June until September, the earth continues its counterclockwise journey around the sun. On September 22 it is 90° from its original position in June. It is still tilted in exactly the same direction. The sun can now reach both the North and South Poles. From the earth, the sun appears exactly over the Equator.

By December 22 the earth's position is exactly opposite to what it was on June 22. The North Pole is now tilted away from the sun. The South Pole is exposed to the sun. And the Southern Hemisphere obtains more sunlight than the Northern Hemisphere. It is now summertime in Australia. Notice that the sun cannot reach the North Pole during this period. That is why there is total darkness during the winters and sunlight throughout the summers at the poles. During the winter, the sun cannot rise because its rays are cut off from the North Pole by the curve of the earth.

On March 21 the sun is back opposite its position of September 22. Both poles are lighted by the sun, which appears directly over the Equator.

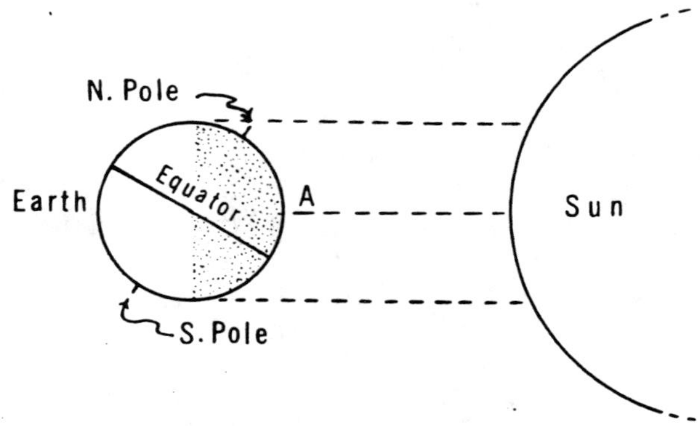

You can see this better in the figure above, which shows the relation between the sun and the earth on June 22. Notice that the sun's rays cannot reach the South Pole because of the curve of the earth. Notice also that point A, where the sun's rays strike directly on the earth, is far north of the Equator. At point A the sun would appear to be directly overhead. If you try to picture the tilted earth revolving around the sun, it will help you to see why there is no sunlight at the poles during the wintertime.

You can prove that the tilt of the earth causes the seasons by using an orange to represent the earth and a piece of wire for its axis. Stick the wire through the two poles and move the orange around an electric light bulb as though it were making an orbit around the sun. Watch the shadow caused by the electric light. You will find that you can see exactly how the seasons work by this simple experiment.

One View of the Earth

In this chapter, we have taken a look at the earth as a planet. In the solar system the earth is just one body orbiting around the sun. Because of the distance from the sun and the rotation, the climate on the earth changes from season to season. Because of the rotation of the earth, we have periods of day and night. The movement of the earth serves man as a clock and a calendar.

This is only one way of looking at the earth. We can call it an *overview* or look at the whole solar system. In the next few chapters, we will look at the earth from another standpoint. We will try to learn what the earth is made of and how it changes. We will study the effect of wind and water on the earth's surface. This is a view of the earth that is more familiar to us. But we must remember that the earth is part of the universe. It is also part of the solar system. It must behave according to the rules of the universe. We cannot change the way the earth takes part in the movement of the solar system. The next best thing to do is learn how the earth acts in the solar system and use this knowledge to plan our own work on earth.

GLOSSARY OF LUNAR SCIENCE

A

ALBEDO
Relative brightness. It is the ratio of the amount of electromagnetic radiation reflected by a body to the amount of incident radiation.

ALPHA PARTICLE
A positive particle consisting of 2 protons and 2 neutrons. It is the nucleus of a helium atom.

ANGSTROM UNIT
A unit of length equal to 10^{-10} meters or 10^{-4} microns. It is approximately four-billionths of an inch. In solids, such as salt, iron, aluminum, the distance between atoms is usually a few Angstroms.

APERTURE
A small opening such as a camera shutter through which light rays pass to expose film when the shutter is open.

ATTENUATION
Decrease in intensity usually of such wave phenomena as light or sound.

B

BASALT
A type of dark gray rock formed by solidification of molten material. The rocks of Hawaii are basalts.

BISTATIC RADAR
The electrical properties of the Moon's surface can be measured by studying the characteristics of radio waves reflected from the Moon. If the radio transmitter and receiver are located at the same place, the term monostatic radar is used. If they are located at different places, then bistatic is used. In the study of the Moon with bistatic radar, the transmitter is aboard the CSM and the receiver is on the Earth.

BRECCIA
A coarse-grained rock composed of angular fragments of pre-existing rocks.

BOUNDARY LAYER
The interaction layer between the solar wind bow shock and the magnetopause.

BOW SHOCK
The shock wave produced by the interaction of the solar wind with the Earth's magnetosphere.

C

CARTOGRAPHY
The production and science of accurately scaled maps.

CASSETTE
Photographic film container. Also magnetic tape container.

CISLUNAR
Pertaining to the space between the Earth and Moon or the Moon's orbit.

COLLIMATOR
A device for producing beams of parallel rays of light or other electromagnetic radiation.

COLORIMETRIC
Pertaining to the measurement of the intensities of different colors as of lunar surface materials.

COSMIC RAYS
Streams of very high energy nuclear particles, commonly protons, that bombard the Earth and Moon from all directions.

COSMOLOGY
Study of the character and origin of the universe.

CRATER
A naturally occurring hole. On Earth, a very few craters are formed by meteorites striking the Earth; most are caused by volcanoes. On the Moon, most craters were caused by meteorites. Some lunar craters were apparently formed by volcanic processes. In the formation of lunar craters, large blocks of rock (perhaps as large as several hundred meters across) are thrown great distances from the crater. These large blocks in turn form craters also -- such craters are termed secondary craters.

CROSS-SUN
A direction approximately 90 degrees to the direction of the Sun and related to lunar surface photography.

CROSSTRACK
Perpendicular to the instantaneous direction of a spacecraft's ground track.

CRYSTALLINE ROCKS
Rocks consisting wholly or chiefly of mineral crystals. Such rocks on the Moon are usually formed by cooling from a liquid melt.

D

DIELECTRIC
A material that is an electrical insulator. Most rocks are dielectrics.

DIURNAL
Recurring daily. Diurnal processes on Earth repeat themselves every 24 hours but on the Moon repeat every 28 Earth days. The length of a lunar day is 28 Earth days.

DOPPLER TRACKING
A system for measuring the trajectory of spacecraft from Earth using continuous radio waves and the Doppler effect. An example of the Doppler effect is the change in pitch of a train's whistle and a car's horn on passing an observer. Because of this effect, the frequency of the radio waves received on Earth is changed slightly by the velocity of the spacecraft in exactly the same way that the pitch of a train's whistle is changed by the velocity of the train.

DOWN-SUN
In the direction that is directly away from the Sun and related to lunar surface photography.

E

EARTHSHINE
Illumination of the Moon's surface by sunlight reflected from the Earth. The intensity is many times smaller than that of the direct sunlight.

ECLIPTIC PLANE
The plane defined by the Earth's orbit about the Sun.

EFFLUENT
Any liquid or gas discharged from a spacecraft such as waste water, urine, fuel cell purge products, etc.; also any material discharged from volcanoes.

EGRESS
A verb meaning to exit or to leave. The popularization of this word has been attributed to the great showman P.T. Barnum, who reportedly discovered that a sign marked exit had almost no effect on the large crowds that accumulated in his exhibit area but a sign marked *to egree* led the crowds outdoors. In space terminology, it means simply to leave the spacecraft.

EJECTA
Lunar material thrown out (as resulting from meteoroid impact or volcanic action).

ELECTRON
A small fundamental particle with a unit of negative electrical charge, a very small mass, and a very small diameter. Every atom contains one or more electrons. The proton is the corresponding elementary particle with a unit of positive charge and a mass of 1,837 times as great as the mass of the electron.

EXOSPHERE
The outermost portion of the Earth's or Moon's atmosphere from which gases can escape into outer space.

F

FIELD
A region in which each point has a definite value such as magnetic field.

FIELD OF VIEW
The region *seen* by the camera lens and recorded on the film. The same phrase is applied to such other equipment as radar and radio antennas.

FILLET
Debris (soil) piled against a rock; several scientists have suggested that the volume of the fillet may be directly proportional to the time the rock has been in its present position and to the rock size.

FLUORESCENCE
Emission of radiation at one wavelength in response to the absorption of energy at a different wavelength. Some lunar materials fluoresce. Most do not. The process is identical to that of the familiar fluorescent lamps.

FLUX
The rate of flow per unit area of some quantity such as the flux of cosmic rays or the flux of particles in the solar wind.

FRONT
The more or less linear outer slope of a mountain range that rises above a plain or plateau. In the United States, the Colorado Front Range is a good example.

G

GALACTIC
Pertaining to a galaxy in the universe such as the Milky Way.

GAMMA
A measure of magnetic field strength; the Earth's magnetic field is about 50,000 gamma. The Moon's magnetic field is only a few gamma.

GAMMA-RAY
One of the rays emitted by radioactive substances. Gamma rays are highly penetrating and can traverse several centimeters of lead.

GARDENING
The overturning, reworking, and changing of the lunar surface due to such processes as meteoroid impact, volcanic action, aging and such.

GEGENSCHEIN
A faint light covering a 20 degree field-of-view projected on the celestial sphere about the Sun-Earth vector (as viewed from the dark side of the Earth).

GEOCHEMICAL GROUP
A group of three experiments especially designed to study the chemical composition of the lunar surface remotely from lunar orbit.

GEODESY
Originally, the science of the exact size and shape of the Earth; recently broadened in meaning to include the Moon and other planets.

GEOPHONE
A small device implanted in the lunar surface during the deployment of the ASE to detect vibrations of the Moon from artificial and natural sources.

GEOPHYSICS
Physics of planetary bodies, such as the Earth and Moon, and the surrounding environment; the many branches include gravity, magnetism, heat flow, seismology, space physics, geodesy, meteorology, and sometimes geology.

GNOMON
A rod mounted on a tripod in such a way that it is free to swing in any direction and indicates the local vertical; it gives Sun position and serves as size scale. Color and reflectance scales are provided on the rod and a colorimetric reference is mounted on one leg.

GRADIENT
The rate of change of something with distance. Mathematically, it is the space rate of change of a function. For example, the slope of a mountain is the gradient of the elevation.

I

IMBRIAN AGE
Two methods of measuring age on the Moon are used. One provides the absolute age, in years, and is based on radioactivity. The other gives only relative ages. A very old event on the Moon is that which produced the Imbrium basin. The age of other geologic features can be determined with respect to the Imbrium event.

INGRESS
A verb meaning to enter. It is used in connection with entering the LM. See also EGRESS.

IN SITU
Literally, *in place, in its original position*. For example, taking photographs of a lunar surface rock sample *in situ* (as it lies on the surface).

L

LIMB
The outer edge of the apparent disk of a celestial body, as the Moon or Earth, or a portion of the edge.

LITHOLOGY
The character of a rock formation.

LUNATION
One complete passage of the Moon around its orbit.

M

MANTLE
An intermediate layer of the Moon between the outer layer and the central core.

MARE
A large dark flat area on the lunar surface (Lunar Sea). May be seen with the unaided eye.

MARIA
Plural of mare.

MASCOONS
Large mass concentrations beneath the surface of the Moon. They were discovered only 3 years ago by changes induced by them in the precise orbits of spacecraft about the Moon.

MASS SPECTROMETER
An instrument which distinguishes chemical species in terms of their different isotopic masses.

METEORITE
A solid body that has arrived on the Earth or Moon from outer space. It can range in size from microscopic to many tons. Its composition ranges from that of silicate rocks to metallic iron-nickel.

METRIC PHOTOGRAPHY
Recording of surface topography by means of photography, together with an appropriate network of coordinates, to form the basis of accurate measurements and reference points for precise photographic mapping.

MICROSCOPIC
Of such a size as to be invisible to the unaided eye but readily visible through a microscope.

MINERALOGY
The science of minerals; deals with the study of their atomic structure and their general physical and chemical properties.

MONOPOLE
All known magnets have two poles, one south pole and one north pole. The existence of a single such pole, termed a monopole, has not yet been established but is believed by many physicists to exist on the basis of theoretical studies. Lunar samples have been carefully searched on Earth for the presence of monopoles.

MORPHOLOGY
The external shape of rocks in relation to the development of erosional forms or topographic features.

MOULTON POINT
A theoretical point along the Sun-Earth line located 940,000 statute miles from the Earth at which the sum of all gravitational forces is zero.

N

NADIR
That point on the Earth (or Moon) vertically below the observer.

NAUTICAL MILE
It is 6,280 feet -- 19 percent larger than a *regular* mile.

NEUTRON
An uncharged elementary particle that has a mass nearly equal to that of a proton and is present in all known atomic nuclei except hydrogen.

O

OCCULTATION
The disappearance of a body behind another body of larger apparent size. For example, the occultation of the Sun by the Moon as viewed by an Earth observer to create a solar eclipse.

OZONE
Triatomic oxygen (O_2); found in significant quantities in the Earth's atmosphere.

P

P-10
A gas mixture consisting of 90 percent argon, 9.5 percent carbon dioxide, and 0.5 percent helium used to fill the x-ray detectors of the x-ray fluorescence experiment.

PANORAMA
A series of photographs taken from a point to cover 360 degrees around that point.

PENUMBRA
The part of a shadow in which the light (or other rays such as the solar wind) is only partially masked, in contrast to the umbra in which light is completely masked by the intervening object.

PETROGRAPHY
Systematic description of rocks based on observations in the field (e.g., on the Moon), on returned specimens, and on microscopic work.

PHOTOMULTIPLIER TUBE
An electron tube that produces electrical signals in response to light. In the tube, the signal is amplified to produce a measurable output current from very small quantities of light.

PLASMA
A gas composed of ions, electrons, neutral atoms, and molecules. The interactions between particles is mainly electromagnetic. Although the individual particles are electrically positive or negative, the gas as a whole is neutral.

POSIGRADE
Lunar orbital motion in the direction of lunar rotation.

PRIMORDIAL
Pertaining to the earliest, or original, lunar rocks that were created during the time between the initial and final formation stages of the Moon.

PROTON
The positively charged constituent of atomic nuclei.

R

RADON
Isotopes of a radioactive gaseous element with atomic number 86 and atomic masses of 220 and 222 formed by the radioactive decay of radium.

RAY
Bright material that extends radially from many craters on the Moon, believed to have been formed at the same time as the associated craters were formed by impacting objects from space; usually, but not always, arcs of great circles. They may be several hundred kilometers long.

REGOLITH
The unconsolidated residual material that resides on the solid surface of the Moon (or Earth).

RETROGRADE
Lunar orbital motion opposite the direction of lunar rotation.

RILLE/RILL
A long, narrow valley on the Moon's surface.

RIM
Elevated region around craters and rilles.

S

SAMPLE
Small quantities of lunar soil or rocks that are sufficiently small to return them to Earth. On each mission several different kinds of samples are collected. Contingency sample consists of 1 to 2 pounds of rocks and soil collected very early in the surface operations so that at least some material will have been returned to Earth in the event that the surface activities are halted abruptly and the mission aborted. Documented sample is one that is collected with a full set of photographs to allow positive identification of the sample when returned to Earth with the sample in situ together with a complete verbal description by the astronaut. Comprehensive sample is a documented sample collected over an area of a few yards square.

S-BAND
A range of frequencies used in radar and communications that extends from 1.55 to 5.2 kilomegahertz.

SCARP
A line of cliffs produced by faulting or erosion.

SEISMIC
Related to mechanical vibration within the Earth or Moon resulting from, for example, impact of meteoroids on the surface.

SHOCKED ROCKS
Rocks which have been formed by or subjected to the extremes of temperature and pressure from impacts.

SOLAR WIND
Streams of particles (mostly hydrogen and helium) emanating from and flowing approximately radially outward from the Sun.

SPATIAL
Pertaining to the location of points in three-dimensional space; contrasted with temporal (pertaining to time) locations.

SPECTROMETER
An instrument which separates radiation into energy bands (or, in a mass spectrometer, particles into mass groups) and indicates the relative intensities in each band or group.

SPUR
A ridge of lesser elevation that extends laterally from a mountain or mountain range.

STELLAR
Of or pertaining to stars.

STEREO
A type of photography in which photographs taken of the same area from different angles are combined to produce visible features in three-dimensional relief.

SUPPLEMENTARY SAMPLE STOP
A stop added to a traverse after the stations are numbered. Mission planning continues through launch and the supplementary sample stops are inserted between normal traverse stations.

SUPRATHERMAL
Having energies greater than thermal energy.

SUBSATELLITE
A small unmanned satellite, deployed from the spacecraft while it is in orbit, designed to obtain various types of solar wind, lunar magnetic, and S-band tracking data over an extended period of time.

T

TALUS
Rock debris accumulated at the base of a cliff by erosion of material from higher elevation.

TEMPORAL
Referring to the passage or measurement of time.

TERMINATOR
The line separating the illuminated and the darkened areas of a body such as the Earth or Moon which is not self-luminous.

TERRA
Those portions of the lunar surface other than the maria; the lighter areas of the Moon. They are visible to the unaided eye.

TIDAL
Referring to the very small movement of the surface of the Moon or the Earth due to the gravitational attraction of other planetary bodies. Similar to the oceanic tides, the solid parts of the Earth's crust rise and fall twice daily about three feet. Lunar tides are somewhat larger. The tides of solid bodies are not felt by people but are easily observed with instruments.

TIMELINE
A detailed schedule of astronaut or mission activities indicating the activity and time at which it occurs within the mission.

TOPOGRAPHIC
Pertaining to the accurate graphical description, usually on maps or charts, of the physical features of an area on the Earth or Moon.

TRANSEARTH
During transit from the Moon to the Earth.

TRANSIENT
A short-lived event that does not repeat at regular intervals, often occurring in a system when first turned on and before reaching operating equilibrium. For example, the initial current surge that occurs when an electrical system is energized.

TRANSPONDER
A combined receiver and transmitter whose function is to transmit signals automatically when triggered by a suitable signal. Those used in space are sensitive to radio signals.

U

UMBRA
The dark central portion of the shadow of a large body such as the Earth or Moon. Compare penumbra.

UP-SUN
Into the direction of the Sun and related to lunar surface photography.

URANIUM
One of the heavy metallic elements that are radioactive.

V

VECTOR
A quantity that requires both magnitude and direction for its specification, as velocity, magnetic force field, and gravitational acceleration vectors.

W

WAVELENGTH
The distance between peaks (or minima) of waves such as ocean waves or electromagnetic waves.

X

X-RAY

Electromagnetic radiation of non-nuclear origin within the wavelength interval of 0.1 to 100 Angstroms (between gamma ray and ultraviolet radiation), x-rays are used in medicine to examine teeth, lungs, bones, and other parts of the human body; they also occur naturally.

Z

ZODIACAL LIGHT

A faint glow extending around the entire zodiac but showing most prominently in the neighborhood of the Sun. (It may be seen in the west after twilight and in the east before dawn as a diffuse glow. The glow may be sunlight reflected from a great number of particles of meteoritic size in or near the ecliptic in the planetoid belt.)

ANSWER SHEET

TEST NO. ____________ PART ______ TITLE OF POSITION ______________________________
(AS GIVEN IN EXAMINATION ANNOUNCEMENT - INCLUDE OPTION, IF ANY)

PLACE OF EXAMINATION ______________________________ (CITY OR TOWN) ____________ (STATE) DATE ____________

RATING

USE THE SPECIAL PENCIL. MAKE GLOSSY BLACK MARKS.

1 A B C D E	26 A B C D E	51 A B C D E	76 A B C D E	101 A B C D E
2 A B C D E	27 A B C D E	52 A B C D E	77 A B C D E	102 A B C D E
3 A B C D E	28 A B C D E	53 A B C D E	78 A B C D E	103 A B C D E
4 A B C D E	29 A B C D E	54 A B C D E	79 A B C D E	104 A B C D E
5 A B C D E	30 A B C D E	55 A B C D E	80 A B C D E	105 A B C D E
6 A B C D E	31 A B C D E	56 A B C D E	81 A B C D E	106 A B C D E
7 A B C D E	32 A B C D E	57 A B C D E	82 A B C D E	107 A B C D E
8 A B C D E	33 A B C D E	58 A B C D E	83 A B C D E	108 A B C D E
9 A B C D E	34 A B C D E	59 A B C D E	84 A B C D E	109 A B C D E
10 A B C D E	35 A B C D E	60 A B C D E	85 A B C D E	110 A B C D E

Make only ONE mark for each answer. Additional and stray marks may be counted as mistakes. In making corrections, erase errors COMPLETELY.

11 A B C D E	36 A B C D E	61 A B C D E	86 A B C D E	111 A B C D E
12 A B C D E	37 A B C D E	62 A B C D E	87 A B C D E	112 A B C D E
13 A B C D E	38 A B C D E	63 A B C D E	88 A B C D E	113 A B C D E
14 A B C D E	39 A B C D E	64 A B C D E	89 A B C D E	114 A B C D E
15 A B C D E	40 A B C D E	65 A B C D E	90 A B C D E	115 A B C D E
16 A B C D E	41 A B C D E	66 A B C D E	91 A B C D E	116 A B C D E
17 A B C D E	42 A B C D E	67 A B C D E	92 A B C D E	117 A B C D E
18 A B C D E	43 A B C D E	68 A B C D E	93 A B C D E	118 A B C D E
19 A B C D E	44 A B C D E	69 A B C D E	94 A B C D E	119 A B C D E
20 A B C D E	45 A B C D E	70 A B C D E	95 A B C D E	120 A B C D E
21 A B C D E	46 A B C D E	71 A B C D E	96 A B C D E	121 A B C D E
22 A B C D E	47 A B C D E	72 A B C D E	97 A B C D E	122 A B C D E
23 A B C D E	48 A B C D E	73 A B C D E	98 A B C D E	123 A B C D E
24 A B C D E	49 A B C D E	74 A B C D E	99 A B C D E	124 A B C D E
25 A B C D E	50 A B C D E	75 A B C D E	100 A B C D E	125 A B C D E

ANSWER SHEET

TEST NO. ______________ PART ________ TITLE OF POSITION ______________________________
(AS GIVEN IN EXAMINATION ANNOUNCEMENT - INCLUDE OPTION, IF ANY)

PLACE OF EXAMINATION ______________________________ DATE ______________
(CITY OR TOWN) (STATE)

RATING

USE THE SPECIAL PENCIL. MAKE GLOSSY BLACK MARKS.

A B C D E	A B C D E	A B C D E	A B C D E	A B C D E
1	26	51	76	101
2	27	52	77	102
3	28	53	78	103
4	29	54	79	104
5	30	55	80	105
6	31	56	81	106
7	32	57	82	107
8	33	58	83	108
9	34	59	84	109
10	35	60	85	110

Make only ONE mark for each answer. Additional and stray marks may be counted as mistakes. In making corrections, erase errors COMPLETELY.

A B C D E	A B C D E	A B C D E	A B C D E	A B C D E
11	36	61	86	111
12	37	62	87	112
13	38	63	88	113
14	39	64	89	114
15	40	65	90	115
16	41	66	91	116
17	42	67	92	117
18	43	68	93	118
19	44	69	94	119
20	45	70	95	120
21	46	71	96	121
22	47	72	97	122
23	48	73	98	123
24	49	74	99	124
25	50	75	100	125